2013

Targeting U.S. Technologies

A Trend Analysis of Cleared Industry Reporting

PREFACE

It was not until the mid-19th century that scientific analysis refuted the archaic yet generally accepted belief that foul air caused disease. According to the long-held "miasma" theory, this noxious influence was generalized, and anyone could fall victim to a variety of diseases at any time. Trying to understand, predict, and prevent the spread of disease was nearly useless.

In 1854, John Snow conducted a scientific analysis that traced a particular disease outbreak to one specific London well, and demonstrated that cholera befell only those who drank water from it. The application of careful analysis provided a better understanding of the spread of disease. Similarly, the Defense Security Service (DSS) applies analysis to an ever-growing body of reports to better identify the threat to sensitive or classified information and technology resident in the U.S. cleared industrial base, and provides industry with the information needed to protect against foreign collectors.

Sometimes it can seem as if our adversaries surround us with a pervasive, "miasmatic" atmosphere, in which dangerous influences can seep in from any direction at any time, poison our industrial, military, and economic health, and mysteriously leave our warfighters' lives endangered and our national security weakened. But we know that, while general factors and overall trends affect the ongoing battle to protect U.S. technology, intellectual property, trade secrets, and proprietary information from foreign targeting, in practice every collector's attack comes in the form of an individual approach, contact, probe, request, or stratagem. A dangerous, generalized miasma of foreign opportunism may seem to swirl invisibly around the U.S. cleared industrial base, but on a daily basis cleared contractors, DSS, and their partners face concrete threats with which they must resolutely grapple in a practical and effective manner.

This annual publication, Targeting U.S. Technologies: A Trend Analysis of Cleared Industry Reporting, represents part of the effort to maximize the effectiveness of those endeavoring to maintain our national security. DSS builds on the information contained in reports from industry to develop analytical assessments that articulate the threat to U.S. information and technology resident in cleared industry. As cleared contractors and DSS personnel "in the field" attempt to seal off cleared industry from antagonists' noxious infiltration attempts, they report on their efforts. DSS then collects those reports and evaluates and analyzes them.

This process benefits our national security, warfighters, cleared industry partners, and local communities. The information contained in this report helps employees, companies, and intelligence and law enforcement professionals better understand the nature of the pervasive influences that oppose us, regarding both continuities and fluctuations. Increased awareness of the U.S. technologies being targeted by foreign entities and the methods of operation used to attempt to acquire those technologies can only make us better at identifying and thwarting illicit collection attempts. In fiscal year 2012, our combined efforts produced 657 operations or investigations based on information that industry provided. Over 95 percent of these are still undergoing significant action, with many foreign collectors already identified, isolated, diverted, or otherwise thwarted.

Foreign collectors do not constitute a miasma. Their efforts to infiltrate the cleared industrial base and sap our strength are instead comprehensible and can be thwarted, but only through a team effort among cleared contractors, DSS, and intelligence and law enforcement partners. Constant and better-attuned vigilance, smarter methods and defenses, and increased and improved mutual support can help our cleared industrial base keep our country strong, healthy, and secure.

Stanley L. Sims
DIRECTOR
DEFENSE SECURITY SERVICE

TABLE OF CONTENTS

22

———

REGIONAL
ANALYSIS

58

———

CONCLUSION

62

———

OUTLOOK

64

———

REGIONAL
BREAKDOWN

EXECUTIVE SUMMARY

In fiscal year 2012 (FY12), the total number of industry reports to the Defense Security Service (DSS) concerning attempts by foreign collectors to obtain illegal or unauthorized access to sensitive or classified information and technology resident in the U.S. cleared industrial base continued to rise. In the FY10 version of this publication, the reported year-over-year rise was 50 percent; in the FY11 version, it was 74 percent; in this year's version, it is 60 percent.

In FY12, industry reporting reached a watermark of sorts. For the first time, one region—East Asia and the Pacific—was the origin of half of all reported incidents, an increase from 43 percent of the total in FY11. The number of reported cases DSS ascribed to every other region also rose from FY11 to FY12. However, given the magnitude of the increase in reports attributed to East Asia and the Pacific, the share of the total ascribed to every other region either stayed the same as in FY11 or declined.

The Near East remained the second most active collector region, based on industry reporting, at 16 percent of the FY12 total. The only change in the relative positions of the different regions with regard to industry reporting of collection attempts linked to associated entities was South and Central Asia's assumption of the third spot at 12 percent, displacing Europe and Eurasia at nine percent.

The even greater predominance of East Asia and the Pacific within the regional attributions meant even greater predominance in the data for suspicious network activity (SNA) as a method of operation (MO), because industry reporting reflects significant reliance on this method by East Asia and the Pacific collectors. This year, SNA became the most commonly reported MO, displacing both attempted acquisition of technology (AAT) and request for information (RFI) from last year. In FY12, AAT stepped back to second position, but RFI slipped to fourth, as academic solicitation, the most commonly reported MO in Near East-connected reports, replaced it in the top three. These changes meant that more direct and transparent methods were increasingly displaced by more indirect and opaque methods.

The predominance of entities linked to East Asia and the Pacific and the Near East in industry reporting had a parallel effect regarding collector affiliations. While commercial collectors remained most common in the data relating to all collectors world-wide, the portion DSS attributed to this category fell from 34 percent in FY11 to 29 percent in FY12. In contrast, government (the most common affiliation in East Asia and the Pacific-connected reports) and government-affiliated (the most common affiliation in Near East-connected reports) both increased their share of the total, from 17 to 25 and from 18 to 22 percent, respectively.

The technologies that industry reported as targeted in collection attempts showed significant stability. The top four technologies were information systems; electronics; lasers, optics, and sensors; and aeronautics systems technologies. They accounted for from eight to 15 percent apiece in FY11 data, and remained the top four in FY12 (albeit in somewhat different order), accounting for from nine to 11 percent of the data. Similarly, the next eight technologies in order of frequency of reporting remained the same between the two years. While there was some repositioning between these individual technologies, and while collectively the eight accounted for over a quarter of the total, each one accounted for only two to five percent of the total, so the variations had little effect on the overall picture. That overall pattern continued to show foreign entities attempting illicit collections against a wide range of technologies.

The most significant movement within the targeted technology data was the surge in reported collection attempts against electronics. The number of cases related to this category increased by 94 percent, significantly higher than the overall increase in reported cases, and the category went from accounting for eight percent of the total in FY11 to 11 percent in FY12. A substantial number of East Asia and the Pacific entities' requests for electronics technology targeted radiation-hardened (rad-hard) integrated circuits, the special focus area of last year's version of this publication. Rad-hard circuits have applications in nuclear weapons, ballistic missiles, aerospace vehicles, and space programs.

Foreign entities, especially those linked to countries with mature missile programs, increasingly focused collection efforts on U.S. missile technology, usually aimed at particular missile subsystems. This version of Targeting U.S. Technologies: A Trend Analysis of Cleared Industry Reporting covers missiles in a special focus area.

KEY FINDINGS

EAST ASIA AND THE PACIFIC

Recent changes in political leadership have caused a number of regimes to reassess or modify their defense postures, approaches, and responsibilities, affecting their collection strategies. More so than most, regimes in the region must address both short- and long-term threats and defense challenges. All this makes the region's record of attempted collections a mix of continuity and change.

East Asia and the Pacific reached 50 percent in industry reports attributed to the region's collectors. While other regions remained active collectors, East Asia and the Pacific's nearest competitor accounted for only about one-third as many reported collection attempts.

Electronics became East Asia and the Pacific's most commonly reported targeted technology, with attempts aimed at specific sensitive components rather than complete systems. But as a variety of end uses for the technology was possible, their ultimate purposes remained unclear.

East Asia and the Pacific cyber actors conducted relentless collection efforts, continually innovated, and demonstrated increasingly sophisticated abilities and targeting. SNA became their foremost reported MO: the number of relevant SNA incidents increased by almost three and a half times, and the category increased its proportion of the total from 23 to 42 percent.

THE NEAR EAST

Significant frictions and international hostility within the Near East and with states from other regions as well as world organizations provided national regimes with a great deal of motivation to buttress defense forces.

International financial strictures and current national economic difficulties provided additional pressures to maintain and improve defense forces

while expending the fewest resources possible, including by enabling cheaper indigenous manufacture.

Near East entities, most often government-affiliated, attempted to exploit delegation visits to cleared contractor facilities when possible, but otherwise tended to use complicated and opaque procurement networks to attempt to avoid export-control regulations.

Academic solicitations increased by almost 80 percent over FY11, becoming the top MO. Students sought to conduct postgraduate-level research at U.S. academic centers involved in sensitive or classified research, especially in mechanical and aerospace engineering, materials science, and electrical and computer engineering.

SOUTH AND CENTRAL ASIA

As U.S. foreign policy priorities change, some South and Central Asia states may begin to realign their relationships toward other states outside the region. The new relationships will likely pose a significant threat of transfer of U.S. defense technology, whether acquired legally or illicitly. Shrinking regional defense budgets will potentially increase the illicit targeting of Western, including U.S., technology as multiple states attempt to maintain and modernize their militaries.

The number of cases ascribed to government-affiliated collectors more than tripled, so the latter category accounted for 37 percent of the FY12 total, just higher than the commercial segment at 36 percent.

While AAT remained the most commonly reported MO, academic solicitation increased its share from nine to 25 percent of the total and climbed to the second position. Instances of AAT largely involved commercial entities acting as procurement agents seeking U.S. technology and military equipment, and who identified the military or some other government entity as the intended end user. Academic collectors demonstrated overt interest in U.S. technology in the large volume of requests from government-affiliated university entities to subject matter experts in U.S. industry.

While electronics jumped from being the fourth most targeted technology sector in FY11 industry reporting to the top position in FY12, South and Central Asia entities continued to display a wide-ranging interest in U.S. information and technology. This almost certainly reflected the military modernization efforts for which states in the region perceived a need for Western technologies. Collectors very likely intended to use many of the targeted technologies to upgrade older variants or to conduct reverse-engineering.

EUROPE AND EURASIA

Europe and Eurasia contains some of the most effective economic competitors to U.S. cleared contractors. The region's economic woes and popular pressures constantly exert pressure toward leaner defense budgets, even as its alliances and traditional international responsibilities continue to require relatively large, modern, well-equipped, and quickly deployable military forces. Europe and Eurasia collectors targeted every technology sector in pursuit of these combined goals. Some countries with less robust export-control regimes found themselves used as pass-through sites for illicit technology-collection attempts, whether wittingly or unwittingly.

Although in FY12, Europe and Eurasia slipped to being the fourth-ranked region in reported collection attempts, those attempts increased by 13 percent, and the region contains some of the most skillful—and worrisome—collectors targeting U.S. information and technology.

Commercial entities were the primary reported Europe and Eurasia collectors in FY12, accounting for more than twice the portion of the next most active, which was government-affiliated entities, often research institutions.

AAT and RFI combined to account for 60 percent of all reported suspicious contacts ascribed to this region.

Figure 1: Fiscal Year 2012 Regional Trends

Note: Categories in the legends are listed in order of prevalence based on overall FY12 reporting.

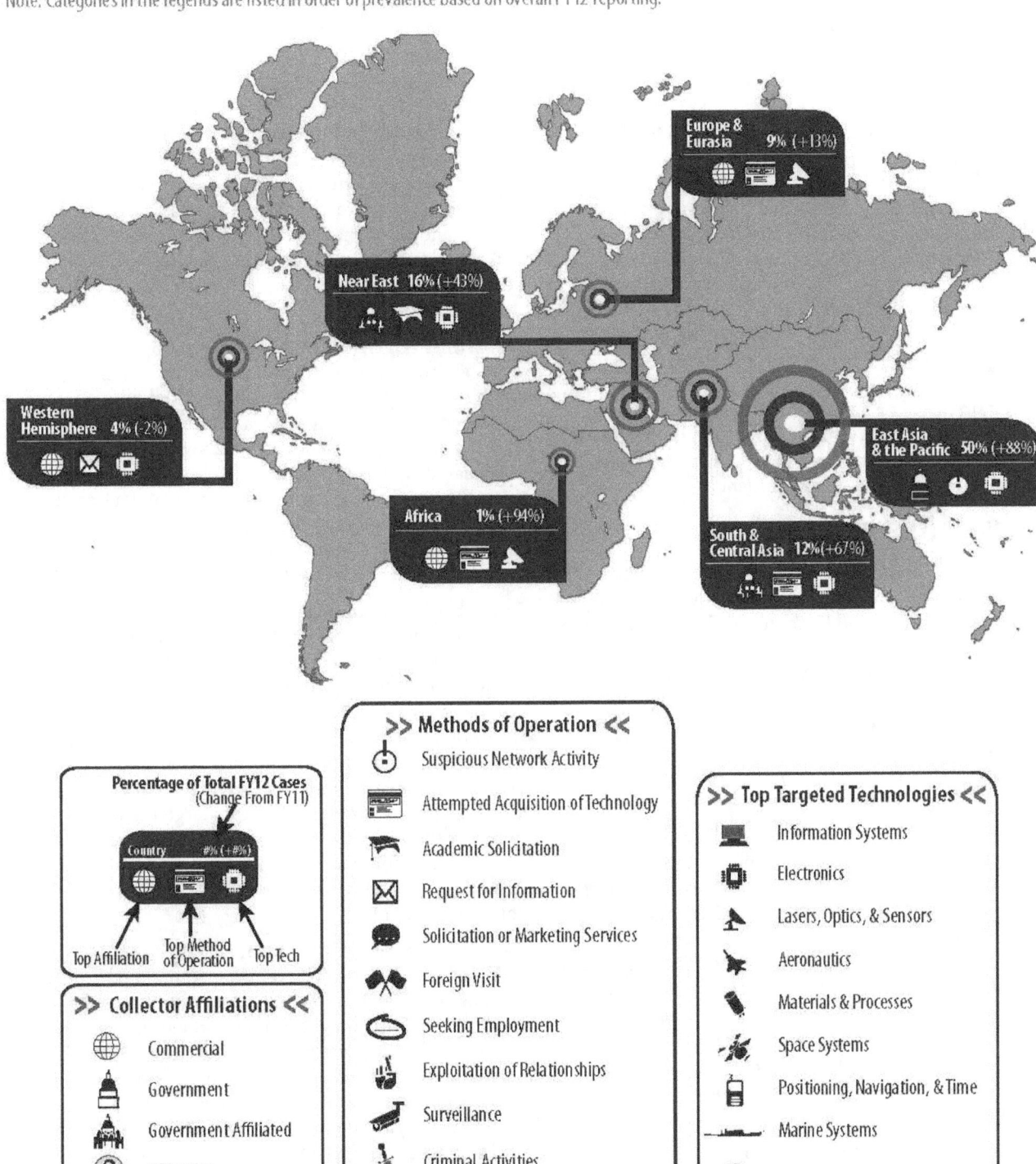

Europe & Eurasia 9% (+13%)

Near East 16% (+43%)

Western Hemisphere 4% (-2%)

Africa 1% (+94%)

East Asia & the Pacific 50% (+88%)

South & Central Asia 12% (+67%)

Percentage of Total FY12 Cases (Change From FY11)

Country #% (+#%)

Top Affiliation   Top Method of Operation   Top Tech

>> Methods of Operation <<

- Suspicious Network Activity
- Attempted Acquisition of Technology
- Academic Solicitation
- Request for Information
- Solicitation or Marketing Services
- Foreign Visit
- Seeking Employment
- Exploitation of Relationships
- Surveillance
- Criminal Activities
- Search/Seizure

>> Collector Affiliations <<

- Commercial
- Government
- Government Affiliated
- Unknown
- Individual

>> Top Targeted Technologies <<

- Information Systems
- Electronics
- Lasers, Optics, & Sensors
- Aeronautics
- Materials & Processes
- Space Systems
- Positioning, Navigation, & Time
- Marine Systems
- Information Security
- Processing & Manufacturing

BACKGROUND

THE CHARGE TO THE DEFENSE SECURITY SERVICE

The Defense Security Service (DSS) supports national security and the warfighter, secures the nation's technological base, and oversees the protection of U.S. and foreign classified information in the hands of industry. The DSS Counterintelligence (CI) Directorate seeks to identify unlawful penetrators of cleared U.S. industry and stop foreign collection attempts to obtain illegal or unauthorized access to classified information and technology resident in the U.S. cleared industrial base. DSS CI articulates the threat for industry and U.S. government leaders.

THE ROLE OF INDUSTRY

In carrying out its mission, DSS relies on the support of cleared contractor employees and the U.S. intelligence and law enforcement communities. Chapter 1, Section 3 of the National Industrial Security Program Operating Manual (NISPOM), 5220.22-M, dated February 28, 2006, requires cleared contractors to remain vigilant and report any suspicious contacts to DSS. The process that begins with initial reporting and continues with ongoing and collective analysis reaches its ultimate stage in successful investigations or operations.

In accordance with the reporting requirements laid out in the NISPOM, DSS receives and analyzes reports from cleared contractors. DSS categorizes these reports as suspicious, unsubstantiated, or of no value. For each reported collection attempt, DSS data aggregation and analysis methodologies seek to gather as much information as possible: who instigated the attempt, where it came from, what its aim was, and what methods of collection it used. The analysis of this information forms the basis for this report.

Such cleared contractor reporting provides information concerning actual, probable, or possible espionage, sabotage, terrorism, or subversion activities to DSS and the Federal Bureau of Investigation. When indicated, DSS refers cases of CI concern to its partners in the law enforcement and intelligence communities for potential exploitation or neutralization. DSS follows up with remedial actions for industry to decrease the threat in the future. This builds awareness and understanding of the individual and collective threats and actions and informs our defenses.

THE REPORT

Department of Defense (DoD) Instruction 5200.39, Critical Program Information (CPI) Protection Within the Department of Defense, dated July 16, 2008, requires DSS to publish a classified report that details suspicious contacts occurring within the cleared contractor community. The focus of this report is on efforts to compromise or exploit cleared personnel, or to obtain illegal or unauthorized access to classified information or technologies resident in the U.S. cleared industrial base.

Each year DSS publishes Targeting U.S. Technologies: A Trend Analysis of Cleared Industry Reporting (formerly carrying the subtitle A Trend Analysis of Reporting from Defense Industry). In this report, the 15th annual Targeting U.S. Technologies (or Trends), DSS provides a snapshot of its findings on foreign collection attempts. It provides a statistical and trend analysis that covers the most prolific foreign collectors targeting the cleared contractor community during fiscal year 2012 (FY12), compares that information to the previous year's report, and places that comparison into a larger context.

DoD Instruction 5200.39 requires DSS to provide its reports to the DoD CI community, national entities, and the cleared contractor community. This report constitutes part of DSS' ongoing effort to assist in better protecting the U.S. cleared industrial base by raising general threat awareness, encouraging the reporting of incidents as they occur, identifying specific technologies at risk, and applying appropriate countermeasures. DSS intends the report to be a ready reference tool for security professionals in their efforts to detect, deter, mitigate, or neutralize the effects of foreign targeting. DSS previously released a classified version of this report.

SCOPE/METHODOLOGY

DSS bases this report primarily on SCRs collected from the cleared contractor community. It also includes references to all-source Intelligence Community (IC) reporting.

DSS considers all SCRs received from cleared industry. It then applies analytical processes to them, including the DSS foreign intelligence threat assessment methodology (explained in the next section). This publication is organized first by collector region, then by collector affiliation, methodologies employed, and technologies, including the specific technology sectors targeted. It incorporates statistical and trend analyses on each of these areas. Each section also contains a forecast of future activities against the cleared contractor community, based on analytical assessments.

DSS continues to analyze foreign interest in U.S. defense technology in terms of the 20 sections of the Militarily Critical Technologies List (MCTL). The MCTL is a compendium of the science and technology capabilities under development worldwide that have the potential to significantly enhance or degrade U.S. military capabilities in the future. It provides categories and subcategories for DSS to use in identifying and defining targeted technologies.

Figure 2: Methods of Operation

Suspicious Network Activity
Via cyber intrusion, viruses, malware, backdoor attacks, acquisition of user names and passwords, and similar targeting, these are attempts to carry out intrustions into cleared contractor networks and exfiltrate protected information

Attempted Acquisition of Technology
Via direct purchase of firms or agency of front companies or third countries, these are attempts to acquire protected information in the form of controlled technologies, whether the equipment itself or diagrams, schematics, plans, spec sheets, or the like

Academic Solicitation
Via requests for, or arrangement of, peer or scientific board reviews of academic papers or presentations, or requests to study or consult with faculty members, or applications for admission into academic institutions, departments, majors, or programs, as faculty members, students, fellows, or employees

Request for Information
Via phone, email, or webcard approaches, these are attempts to collect protected information under the guise of price quotes, marketing surveys, or other direct and indirect efforts

Solicitation or Marketing Services
Via sales, representation, or agency offers, or response to tenders for technical or business services, these are attempts by foreign entities to establish a connection with a cleared contractor vulnerable to the extraction of protected information

Foreign Visit
Via visits to cleared contractor facilities that are either pre-arranged by foreign contingents or unannounced, these are attempts to gain access to and collect protected information that goes beyond that permitted and intended for sharing

Seeking Employment
Via résumé submissions, applications, and references, these are attempts to introduce persons who, wittingly or unwittingly, would thereby gain access to protected information that could prove useful to agencies of a foreign government

Exploitation of Relationships
Via establishing connections such as joint ventures, official agreements, foreign military sales, business arrangements, or cultural commonality, these are attempts to play upon existing legitimate or ostensibly innocuous relationships to gain unauthorized access

Surveillance
Via visual, aural, electronic, photographic, or other means, this comprises systematic observation of equipment, facilities, sites, or personnel

Criminal Activities
Via theft, these are attempts to acquire protected information with no pretense or plausibility of legitimate acquisition

Search/Seizure
Via physical searches of persons, environs, or property or otherwise tampering therewith, this involves temporarily taking from or permanently dispossessing someone of property or restricting his/her freedom of movement

Figure 3: Collector Affiliations

Commercial
Entities whose span of business includes the defense sector

Government
Ministries of Defense and branches of the military, as well as foreign military attachés, foreign liaison officers, and the like

Government Affiliated
Research institutes, laboratories, universities, or contractors funded by, representing, or otherwise operating in cooperation with a foreign government agency, whose shared purposes may include acquiring access to U.S. sensitive, classified, or export-controlled information

Unknown
Instances in which no attribution of a contact to a specific end user could be directly made

Individual
Persons who, for financial gain or ostensibly for academic or research purposes, seek to acquire access to U.S. sensitive, classified, or export-controlled information or technology, or the means of transferring it out of the country

This publication also makes reference to the Department of Commerce's Entity List. This list provides public notice that certain exports, re-exports, and transfers (in-country) to entities included on the Entity List require a license from the Bureau of Industry and Security. An End-User Review Committee (ERC) annually examines and makes changes to the list, as required. The ERC includes representatives from the Departments of Commerce, Defense, Energy, State, and, when appropriate, Treasury.

For FY12, the categories DSS used to identify methods of operation underwent some modification from the previous year. Official Foreign Travel became simply Foreign Visit. Conferences, Conventions, and Trade Shows and Targeting U.S. Travelers Overseas were dropped, and Search and Seizure as well as Surveillance were added.

ESTIMATIVE LANGUAGE AND ANALYTIC CONFIDENCE

DSS employs the IC estimative language standard. The phrases used, such as *we judge, we assess,* or *we estimate,* and terms such as *likely* or *indicate* represent the agency's effort to convey a particular analytical assessment or judgment.

Because DSS bases these assessments on incomplete and at times fragmentary information, they do not constitute facts nor provide proof, nor do they represent empirically based certainty or knowledge. Some analytical judgments are based directly on collected information, others rest on previous judgments, and both types serve as building blocks. In either variety of judgment, the agency may not have evidence showing something to be a fact or that definitively links two items or issues.

Intelligence judgments pertaining to likelihood are intended to reflect the approximate level of probability of a development, event, or trend. Assigning precise numerical ratings to such judgments would imply more rigor than the agency intends. The chart below provides a depiction of the relationship of terms to each other.

Remote	Very Unlikely	Unlikely	Even Chance	Probably, Likely	Very Likely	Almost Certainly

The report uses *probably* and *likely* to indicate that there is a greater than even chance of an event happening. However, even when the authors use terms such as *remote* and *unlikely,* they do not intend to imply that an event will not happen. The report uses phrases such as *we cannot dismiss, we cannot rule out,* and *we cannot discount* to reflect that, while some events are unlikely or even remote, their consequences would be such that they warrant mentioning.

DSS uses words such as *may* and *suggest* to reflect situations in which DSS is unable to assess the likelihood of an event, generally because relevant information is sketchy, fragmented, or nonexistent.

In addition to using words within a judgment to convey degrees of likelihood, DSS also assigns analytic confidence levels based on the scope and quality of information supporting DSS judgments:

HIGH CONFIDENCE

- Well-corroborated information from proven sources, minimal assumptions, and/or strong logical inferences
- Generally indicates that DSS based judgments on high-quality information, and/or that the nature of the issue made it possible to render a solid judgment

MODERATE CONFIDENCE

- Partially corroborated information from good sources, several assumptions, and/or a mix of strong and weak inferences
- Generally means that the information is credibly sourced and plausible but not of sufficient quality or corroborated sufficiently to warrant a higher level of confidence

LOW CONFIDENCE

- Uncorroborated information from good or marginal sources, many assumptions, and/or mostly weak inferences
- Generally means that the information's credibility or plausibility is questionable, or that the information is too fragmented or poorly corroborated to make solid analytic inferences, or that we have significant concerns or problems with the sources

86% increase in reports targeting **missile technology**

54% of all reports originated in **East Asia & the Pacific**

That's a **164% increase** in reports over last year

Reports of incidents originating from the **Near East** increased by **96%**

Attempts to collect missile technology information during **foreign visit** increased by **458%**

98% increase in reported incidents of attempted acquisition of technology

SPECIAL FOCUS AREA: U.S. MISSILE TECHNOLOGY

OVERVIEW

U.S. cleared contractors lead the world in researching, designing, and engineering offensive and defensive missiles. These missiles support U.S. national security and U.S. interests abroad. Over the past year, reporting from industry showed an increase in foreign entity interest in U.S. missile technology. Based on industry reporting, foreign entities focused collection activities on cleared contractors producing various missile technologies grouped in missile subsystems.

Foreign entities' interest in missile technology has risen over the past several years, as reflected in industry reporting culminating in fiscal year 2012 (FY12). Reported targeting of missile technology rose 86 percent from FY11. Defense Security Service (DSS) analysis of industry reporting revealed that regions composed of countries with mature missile programs showed a particularly strong interest in these missile technologies.

Figure 4: Missile Components

Structural Components
airframe, nose cone, nozzles

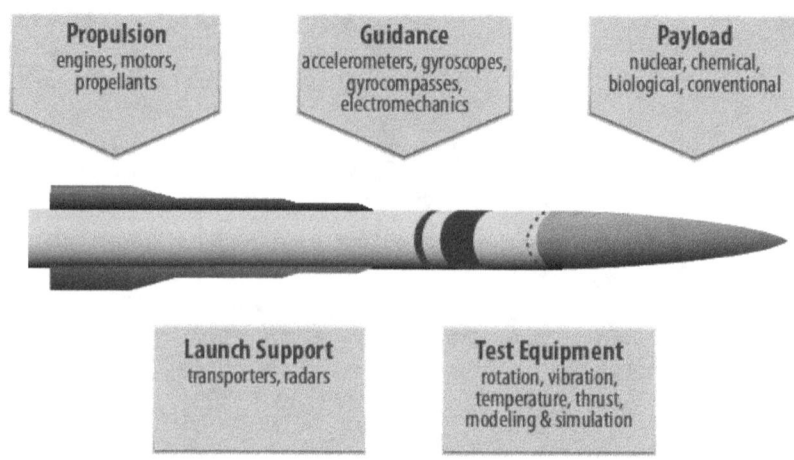

Propulsion
engines, motors, propellants

Guidance
accelerometers, gyroscopes, gyrocompasses, electromechanics

Payload
nuclear, chemical, biological, conventional

Launch Support
transporters, radars

Test Equipment
rotation, vibration, temperature, thrust, modeling & simulation

Although cleared industry reported commercial and government entities attempting to acquire missiles for various end uses, a majority of the reporting showed foreign interest in military and dual-use commodities. A country can use these technologies for missile research, development, manufacturing, and testing and evaluation.

DSS produced this special focus area to alert cleared industry to the increasing foreign threat to U.S. missile technology and facilitate the implementation of mitigation strategies to counter that threat.

After a country masters the chemistry and physics required to launch a missile, scientists and engineers can focus on accuracy and lethality, the desired characteristics of modern missiles. Ultimately, this increases the efficiency and effectiveness of current missiles in a country's order of battle and enhances the country's developmental missile programs. Specialized missile subsystems, such as guidance and test equipment, are crucial to making a missile more accurate and lethal.

From FY11 to FY12, industry reports of foreign collection attempts directed at cleared contractors that research, design, and manufacture missile technology increased by 86 percent. Among these collectors, East Asia and the Pacific-connected entities were the most pervasive in their attempts to collect missile technology, accounting for 54 percent of all reporting received from cleared industry. Collectors linked to the Near East and to Europe and Eurasia completed the top three in reported collection attempts against missile technology in FY12. Although the percentage of Near East entities' attempts to collect missile technology increased only one percent from FY11 to FY12, from 14 to 15 percent of the total, the region's ranking rose to

second from fourth. Together, collectors associated with these three regions accounted for 83 percent of all reported targeting.

Based on industry reporting, foreign entities relied on four methods of operation (MOs) when targeting U.S. missile technology designers, engineers, and manufacturers: attempted acquisition of technology (AAT); foreign visit; request for information (RFI); and solicitation or marketing. These MOs accounted for 75 percent of FY12 reported collection attempts against missile-related technology.

> ### What is a missile?
>
> Merriam-Webster defines a missile as an object thrown or projected, usually so as to strike a target. The National Nuclear Security Administration (NNSA) defines a missile as an airborne delivery vehicle for conventional and unconventional munitions. Unconventional missiles carry nuclear, biological, or chemical payloads. The NNSA definition goes on to say that missiles are unmanned and have a powered flight phase. Thus, conventional airplanes are not missiles, but unmanned aerial vehicles are. The United States and countries throughout the world attempt to control the proliferation of missile technology through various regulations and agreements: International Traffic in Arms Regulations, the Wasenaar Agreement, the Missile Technology Control Regime, Export Administration Regulations, and the Commodities Control List.
>
> Generally, a missile's purpose defines its type, of which there are five: surface-to-surface, air-to-surface, surface-to-air, air-to-air, and anti-satellite. Missile technology is categorized into six major missile subsystems: payload, structural components, guidance, propulsion, test equipment, and launch support.

In data analyzed for this report, suspicious network activity (SNA) does not emerge as a preferred MO used by foreign entities—including those from East Asia and the Pacific—when targeting the broad category of U.S. missile technology. Two reasons likely explain the low incidence of reporting of this MO. First, even when SNA actors exploit vulnerabilities in networks at facilities, it does not always result in the compromise or exfiltration of data. In order to remain consistent in the methodologies used to analyze the data set, it includes only successful exfiltrations of missile-related technology. This methodological necessity does not negate the severity of the SNA MO in targeting missile-related technology, but the number of such successful instances is relatively low across the data set. Second, when SNA goes unrecognized or unreported by cleared contractors, industry does not generate a report, making such instances unavailable for analysis in this data set. (See a related discussion of these difficulties in the East Asia and the Pacific section of this report.)

REGIONS OF ORIGIN

EAST ASIA AND THE PACIFIC

Some of the countries having the world's most mature and active missile research and development programs, outside of the United States, are in East Asia and the Pacific. East Asia and the Pacific entities accounted for 54 percent of FY12 industry reporting on the targeting of missile technology. Their collection focus was predominantly guidance subsystems.

While some countries' research and development focuses on offensive missiles, most missile development in this region focuses on defensive missile production. Specifically, countries are seeking missile technology required to defeat both air-breathing and non-air-breathing threats.

Analyst Comment: The scope of East Asia and the Pacific missile modernization programs encompasses all the five main missile types. East Asia and the Pacific regimes targeting U.S. missile technology likely intend any illicitly acquired information and technology to support their ongoing missile modernizations. (Confidence Level: Moderate)

NEAR EAST

At 15 percent of the FY12 total, industry reporting linked to Near East entities was second in number of reported attempts to collect U.S. missile information and technology. As with East Asia and the Pacific, missile technology categorized in the guidance subsystem was the focus of Near East entities' reported collection attempts.

Tactical and ballistic missile inventories in this region are large, but most national inventories lack the technical sophistication of U.S. and East Asia and the Pacific missiles to hit a target accurately. To compensate for the lack of accuracy, countries rely on quantity of offensive and defensive systems to effect the intended battle damage.

Analyst Comment: Near East countries are likely attempting to collect U.S. missile guidance technology to increase the accuracy and effectiveness of missiles currently in their inventories. Modernization of these systems would likely support future missile development that would focus on the range and accuracy of offensive and defensive missiles. (Confidence Level: High)

Case Study

On February 22, 2012, an Alabama-based cleared contractor received an unsolicited email from an individual representing a Near East defense contractor. The individual requested high-temperature measurements of four-point bending, as defined by the Active Standard Test Method (ASTM) C 1211 for advanced ceramics.

According to its website, the Near East defense contractor develops and manufactures a range of advanced military systems, including naval, air, and ground precision weapons; electro-optics; electronic warfare systems; command, control, communications, computers, and intelligence (i.e., C4I) equipment; acoustic defenses; training and simulation systems; and armored protection.

ASTM C 1211 is a recommended test method for ceramics that will meet MIL STD 1942 (A), Flexural Strength of High Performance Ceramics at Ambient Temperatures. This test method is used for material development, quality control, characterization, and design data generation purposes. According to ASTC International, the method is intended to be used with ceramics whose flexural strength is 50 MPa or greater, which would be suitable for use in missile structures.

According to Intelligence Community reporting, the Near East defense contractor has a well-documented history of attempting to illicitly acquire U.S. technology. Some of these attempts involved U.S. missile technology.

Analyst Comment: Given the history of individuals representing this Near East defense contractor attempting to illicitly collect U.S. missile technology, combined with the company's line of business, this individual likely sought the advanced ceramic test data for a military purpose. Although the individual did not specify an end use, the ceramic test data, if acquired, likely would have been used to develop missile structures more capable of surviving high-temperature environments. (Confidence Level: Moderate)

EUROPE AND EURASIA

In FY12, Europe and Eurasia-connected entities' share of reported collection activities targeting U.S. missile technology fell from 18 percent in FY11 to 14 percent. However, Europe and Eurasia entities remained among the top three reported collectors of U.S. missile technology. Consistent with reported East Asia and the Pacific and Near East collection attempts against missile technology, Europe and Eurasia entities also focused collection efforts on the guidance subsystem category.

Some Europe and Eurasia countries have strong alliances with the United States, and some will host U.S. anti-ballistic missile systems. In addition, most European countries are signatories to the Missile Technology Control Regime, a voluntary association aimed at nonproliferation of unmanned delivery systems. However, some countries and companies incorporated in Europe and Eurasia

continue to seek missile technology to support missile research and the modernization of current missile inventories, and to sell missile technology on the global arms market.

Analyst Comment: Although Europe and Eurasia countries tend to be strongly allied to the United States, most entities within this region likely have some interest in acquiring U.S. missile technology. Acquisition of U.S. missile technology would probably fulfill two goals for these countries. First, acquisition of guidance technology would likely

Figure 5: Primary Regions Targeting U.S. Missile Technology

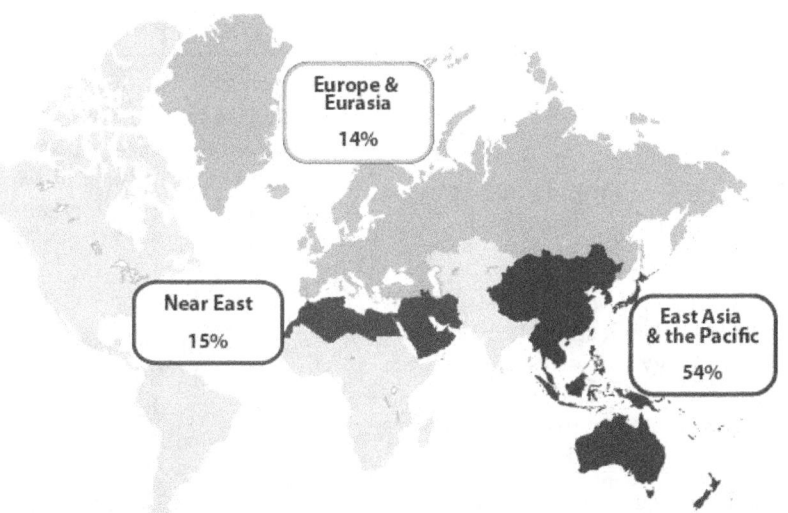

East Asia & Pacific requests spanned four missile subsystems
Payload: Fuses
Structural Components: Composite materials
Guidance: Lasers; accelerometers; traveling wave tubes; monolithic microwave integrated circuits; global positioning system source code; inertial navigation systems; rad-hard microelectronics; infrared and optical receivers; avionics software; magnetrons; precision-guidance kits; data links; target-tracking algorithms; focal plane arrays
Propulsion: Rocket-based combined-cycle propulsion; propellant-actuated devices; hydrogen burnoff devices
Test Equipment: Missile modeling software; electro-optic test sets; intent specification software

Near East requests spanned four missile subsystems
Structural Components: High-temperature ceramics; nanotechnology
Guidance: Lasers; rad-hard microelectronics; optics; miniature vertical gyroscopes; avionics software; traveling wave tubes; micro-electromechanical systems
Propulsion: Rocket motors
Test Equipment: Harpoon missile test equipment and infrastructure; microelectronic failure and vulnerability analysis; missile targeting test set

Europe & Eurasia requests spanned three missile subsystems
Structural Components: Composite materials
Guidance: Rad-hard microelectronics; ruggedized power supplies; active radar; global positioning systems; klystron tubes; selective availability anti-spoofing modules
Test Equipment: modeling and simulation software

support missile research and the modernization of tactical and ballistic missiles in current inventories. Second, the acquisition of missile technology would likely enable companies and government organizations within Europe and Eurasia to effect proliferation through various technology-sharing agreements or lax controls on missile-related technology, for either political or profit-driven reasons. (Confidence Level: Moderate)

AFFILIATIONS AND METHODS OF OPERATION

In analyzing FY12 reports from industry regarding foreign collection attempts to obtain illegal or unauthorized access to sensitive or classified information and technology resident in the U.S. cleared industrial base, once DSS established the entity's country of origin, it then identified its affiliation and the MO used. The following paragraphs detail the top affiliations and MOs identified in FY12 reporting from cleared industry related to missiles.

DSS analysis of industry reporting showed that the collectors most active in collection efforts aimed at U.S. missile technology—those associated with East Asia and the Pacific—primarily relied on commercial entities to attempt to obtain sensitive or classified U.S. technology in FY12.

These entities did so using a variety of MOs. At 27 percent of reports, the attempted acquisition of technology MO was used most often, usually consisting of email attempts to circumvent U.S. laws restricting the export of U.S. missile technology. They also relied on foreign visit, RFI, and solicitation or marketing in 18, 15, and 15 percent of reports, respectively.

Near East efforts, as reflected in industry reporting, relied most on government-affiliated entities to attempt to collect missile-related technology, representing 40 percent of reporting from this region. Unlike East Asia and the Pacific entities, which relied on low-cost, high-gain MOs in their attempts to collect missile-related technology, Near East entities used academic solicitation and foreign visit as their primary means to attempt to collect U.S. missile technology. These MOs accounted for 49 percent of reported Near East-connected attempts to collect U.S. missile technology.

Table 1: U.S. Missile Programs Targeted by the Top Three Regions Targeting Missile Technology in FY12

	PROGRAMS TARGETED
East Asia and the Pacific	Standard Missile -3 (SM-3); SM-6; Hawk; Tube-Launched, Optically Tracked, Wire-Guided Missile (TOW); Trident; Tactical Tomahawk; Patriot Advanced Capability-3 (PAC-3); Ground-Based Interceptor (GBI); Javelin; Mk-104; Stinger; Sea Sparrow; Harpoon
Near East	SM-3; Hawk; BQM-74 TOW; PAC-3; GBI; Harpoon
Europe and Eurasia	Hawk; PAC-3; MEADS; Advanced Tactical Missile System; Stinger; GBI; Minuteman; Harpoon

Based on industry reporting, Europe and Eurasia entities relied on AAT and commercial collectors' foreign visits in their attempts to gain sensitive or classified missile information and technology in FY12.

OUTLOOK

Reporting from industry confirms that U.S. missile technology is of interest to several regions. DSS assesses that the trend of East Asia and the Pacific, Near East, and Europe and Eurasia agents attempting to collect U.S. missile technology will likely continue in the future. (Confidence Level: High)

While cleared industry will almost certainly continue to receive requests for entire weapon systems, the focus of the top regions attempting to collect against missile technology will almost certainly continue to be on technology comprising the six major missile subsystems. (Confidence Level: High)

These regions are likely to continue using a variety of MOs, most prominently AAT. Entities falling into all four main collector affiliations—commercial, government-affiliated, government, and individual—will likely attempt to collect U.S. missile information and technology. In addition, given East Asia and the Pacific collectors' prolific use of SNA to target U.S information and technology in general, they will almost certainly use this MO to target U.S. missile technology in the future. (Confidence Level: High)

World leaders in missile research, development, manufacturing, and testing already have established missile infrastructures, which allows them to focus on making their missiles more accurate and lethal. DSS assesses that such countries likely have more to gain than others from the illicit acquisition of U.S. missile technology, by incorporating state-of-the-art U.S. missile technology into their own missiles, then proliferating that technology through technical assistance or military sales. (Confidence Level: Moderate)

If acquired, U.S. missile technology would likely assist East Asia and the Pacific missile modernization in two ways. First, reverse-engineering would probably give East Asia and the Pacific scientists and engineers a better understanding of the capabilities of the targeted and acquired technology to develop countermeasures to U.S. weapon systems. Second, reverse-engineering followed by mass production of the reverse-engineered technology would likely spur missile modernization by incorporation of that technology into East Asia and the Pacific missiles. (Confidence Level: Moderate)

Near East entities will almost certainly continue to attempt to collect U.S. missile technology, likely by using students and foreign visitors at cleared contractor facilities. The priority for Near East collectors will likely be U.S. missile technology information that could enhance the accuracy of offensive and defensive missiles currently stockpiled in Near East countries. (Confidence Level: Moderate)

Entities from Europe and Eurasia will likely continue to pursue U.S. missile technology for exploitation and proliferation on the global arms markets. Also, U.S. missile technology, if acquired, will likely be at risk of exploitation through technology-sharing agreements between countries from Europe and Eurasia and countries from the Near East and East Asia and the Pacific. (Confidence Level: Moderate)

CASE STUDY

Missile modeling and simulation (M&S) software, which falls into the test equipment category among the six major missile subsystems, was a highly sought-after U.S. missile technology in FY12, second only to U.S. missile guidance technology in reported targeting. M&S technology can support nascent missile programs but can also fill technology gaps for mature missile programs. Based on industry reporting, solid rocket propellant performance prediction software was the most sought-after M&S technology in FY12, and South and Central Asia entities were the most prolific in their attempts to collect this software.

Cleared industry reported that during FY12 individuals representing a South and Central Asia government organization carried out numerous collection attempts:

- On two occasions, a cleared contractor received an email from an individual representing a South and Central Asia government organization specializing in space launch systems. The individual requested information on solid propellant performance software and material ablation, conduction, and erosion analysis software.

- On three occasions, the same cleared contractor received an email from another individual representing the same South and Central Asia government organization. The individual's request was identical to the request above for solid propellant performance software and material ablation, conduction, and erosion analysis software.

- Finally, an individual representing a subordinate organization of the same South and Central Asia government organization requested solid propellant performance software from the same cleared contractor.

The solid propellant performance prediction software provides a method to predict the average delivered performance, mass flow, pressure, thrust, and impulse as functions of time. The prediction of solid fuel performance is important for organizations responsible for research and development on space delivery systems and missiles. Typically, an organization will use the same vehicle to deliver any number of satellites or manned vehicles to space.

Given the different masses of the launch vehicle payload, it becomes imperative to be able to predict the performance of the solid propellants using numerous variables. This software has the potential to reduce propulsion research investment and time, lessen test and evaluation costs, and increase the probability of a successful launch. More significantly, this software can influence the accuracy and lethality of a missile by predicting the chemistry of the solid fuel, thrust energy, and impulses required to effectively deliver conventional and unconventional payloads on the intended targets.

Analyst Comment: The government organization requesting the software from cleared industry would likely use the software, if acquired, to realize efficiencies in solid propellant use for space systems; however, the software could also be used to model and predict solid propellant performance in conventional and unconventional missiles as well. (Confidence Level: High)

EAST ASIA AND THE PACIFIC

50
Percentage of all reported FY12 incidents that originated in East Asia & the Pacific

that is an **88%** increase

102% increase in the number of reported collection attempts against electronics

245% increase in reported incidents of suspicious network activity

198% increase in solicitation or marketing services

42% increase in academic solicitation

43% increase in reports of attempted acquisition of technology

204% increase in reported cases tied to **government** collectors

Reports involving **government-affiliated** collectors rose by **103%**

77% increase in the number of reports targeting **aeronautics systems**

OVERVIEW

As noted each year since the Defense Security Service (DSS) started compiling its annual Targeting U.S. Technologies report (Trends), in fiscal year 2012 (FY12) collection efforts linked to East Asia and the Pacific represented the most significant and prolific threat against information and technology resident in cleared industry, based on industry reporting. Suspicious incidents reported by cleared industry and connected to East Asia and the Pacific increased by 88 percent over FY11. Requests originating in or assessed as affiliated with East Asia and the Pacific accounted for half of all industry reporting DSS received in FY12, an increase from 43 percent the year before.

East Asia and the Pacific as a region has become the subject of much world and U.S. attention. It is viewed as an area of great potential, but also as a potential threat to others. Yet within the region, many states believe they are threatened, both by regional and extra-regional actors. This situation of both possibility and uncertainty leaves many regimes concerned for their ability to defend their countries from various dangers in various scenarios.

That being said, the region contains countries of great disparity: in size, both geographically and economically; in dynamism, both economically and policy-wise; in ambition, both economically and geopolitically; in closeness of relations with the United States, especially regarding degree of defense cooperation; and in willingness to attempt to obtain illegal or unauthorized access to sensitive or classified information and technology resident in the U.S. cleared industrial base.

A number of regimes within East Asia and the Pacific have recently undergone or are undergoing changes in their political leadership. A number are in the midst of reassessing and/or modifying their defense postures, approaches, and responsibilities, with concomitant effects on their collection strategies. In general, regimes in the region must address both short- and long-term threats and defense challenges. All this makes the region's record of attempted collections a mix of continuity and change.

It is continuity that reigns in the listing of technologies that East Asia and the Pacific collectors targeted, especially at the top. Based on industry reporting, the three most commonly targeted technology sectors in FY12 were electronics; information systems (IS); and lasers, optics, and sensors (LO&S) technologies. These three top technologies, albeit in slightly different order, were the same in the FY11 listing for East Asia and the

Pacific. Aeronautics was unchanged in position at fourth. These four technology sectors were also the same as those in the overall world listing for FY12, again in somewhat different order.

Within this continuity in East Asia and the Pacific-linked industry reporting, the most significant shift from FY11 to FY12 was the rise of electronics systems as the most commonly targeted technology area, from third most common at eight percent of the total in FY11 to ten percent of the much larger total in FY12, reflecting a doubling in number of cases. The prominence of reports related to electronics systems displaced IS, the most commonly reported targeted technology category in FY11 results. Regarding electronics, East Asia and the Pacific collectors primarily focused on obtaining specific sensitive components, whereas targeting of complete systems was rare. The sought-after components often had a variety of potential end uses, making their ultimate purpose unclear.

But East Asia and the Pacific-initiated requests for technologies once again covered nearly every category of the Militarily Critical Technologies List (MCTL) in FY12. The technologies in positions five through 12 were the same in FY12 as in FY11. While their relative positions changed in many cases, none of these technologies accounted for more than four percent of the total in FY12.

East Asia and the Pacific collectors relied on suspicious network activity (SNA) as their foremost method of operation (MO) for attempting to obtain illegal or unauthorized access to sensitive or classified information resident in the U.S. cleared industrial base, even more so in FY12 than in FY11. The number of SNA incidents linked to East Asia and the Pacific increased by almost three and a half times, and the category increased its proportion of the total from 23 to 42 percent.

East Asia and the Pacific cyber actors constitute a serious danger because of their relentless targeting and continual innovation. Their tactics have matured with each passing year, and they increasingly demonstrate an ability to combine sophisticated technical abilities with sound intelligence targeting practices. Their seasoned cyber tradecraft has led to multiple intrusions of unclassified networks in cleared industry, defeating some of the most advanced computer network defenses in the world.

The expansion of East Asia and the Pacific cyber operations is readily observable, including in those directed against U.S. cleared industry. This growth is evident in the 1,443 percent increase in reported East Asia and the Pacific-attributed SNA between fiscal year 2009 (FY09) and FY12. In FY12, East Asia and the Pacific cyber entities were responsible for 40 percent of all East Asia and the Pacific-attributed industry reporting and 64 percent of all cyber reporting DSS collected. Industry reporting to DSS also reflected East Asia and the Pacific cyber actors' improved technical capabilities. From FY09 to FY12, the number of confirmed East Asia and the Pacific intrusions into cleared industry's unclassified networks grew by 1,138 percent. DSS combined FY12 industry reporting with reporting aggregated from government partners that identified a cleared contractor as the target of SNA.

Based on industry reporting, East Asia and the Pacific collectors' next most preferred MOs were attempted acquisition of technology (AAT), academic solicitation, solicitation or marketing, and request for information (RFI), each of which accounted for 16 to ten percent of the FY12 total. Foreign visits accounted for another five percent. Of these, only solicitation or marketing increased in both number of cases and percentage of the total; all others increased in number of cases but declined or did not change in percentage of the total, given the marked increase in SNA.

In FY12 industry reporting, government entities were the most common East Asia and the Pacific collectors, at 41 percent of the total, up from 25 percent in FY11, taking over from commercial as the top affiliation. This was largely driven by DSS ascribing most SNA to government entities. Commercial and government-affiliated entities became the next most common reported affiliations. Both increased in number of reported cases, by 68 and 103 percent, respectively. The share of the total accounted for by the commercial affiliation decreased from 31 to 28 percent, while that of government-affiliated increased from 15 to 17 percent.

COLLECTOR AFFILIATIONS

As previously mentioned, government was the affiliation most commonly noted in industry reporting related to East Asia and the Pacific in FY12. The government category accounted for 41 percent of the total, followed by commercial and government-affiliated. Combined, government, commercial, and government-affiliated entities accounted for 86 percent of reported East Asia and the Pacific collection attempts.

In some cases these combined efforts were tightly interwoven, particularly in matters relating to technology collection and defense sales, involving a coordinated approach, using collectors of multiple affiliations and MOs in concert. In other cases, collection activity did not necessarily constitute a single, coordinated effort but instead appeared to be a collection of separate but related efforts. Collection patterns constantly evolve as collection entities incorporate both new requirements and new capabilities and make adjustments based on the success of previous collection efforts.

In an instance of the coordinated approach, during a three-month period, representatives of a cleared contractor had numerous face-to-face meetings with different East Asia and the Pacific entities interested in radar technology. The cleared contractor first hosted a delegation of government researchers, followed by cleared contractor visits to the facilities of and meetings with representatives of separate East Asia and the Pacific commercial firms as well as a government entity and a government-affiliated research institute.

Analyst Comment: Many activities that DSS assessed to be East Asia and the Pacific-connected collection attempts involved collectors with different affiliations targeting identical technologies. In these cases it is likely that the collection requirements were the result of and were addressed pursuant to a coordinated national strategy, involving government, industry, and academia. (Confidence Level: Moderate)

The number of cases associated with commercial entities increased by two-thirds in FY12, although the percentage of such cases actually experienced a small decrease from FY11. Many of these instances involved East Asia and the Pacific companies contacting cleared contractors, usually via email, practicing AAT or RFI through overt means or trying to form a business relationship with the cleared contractor. The majority of these requests involved companies that did not act in an illicit manner.

However, many commercial entities failed to provide any information regarding end use or end user, even after the cleared contractor asked for it. It was also significant that many different commercial entities often requested the same technologies

Figure 6: Collector Affiliations

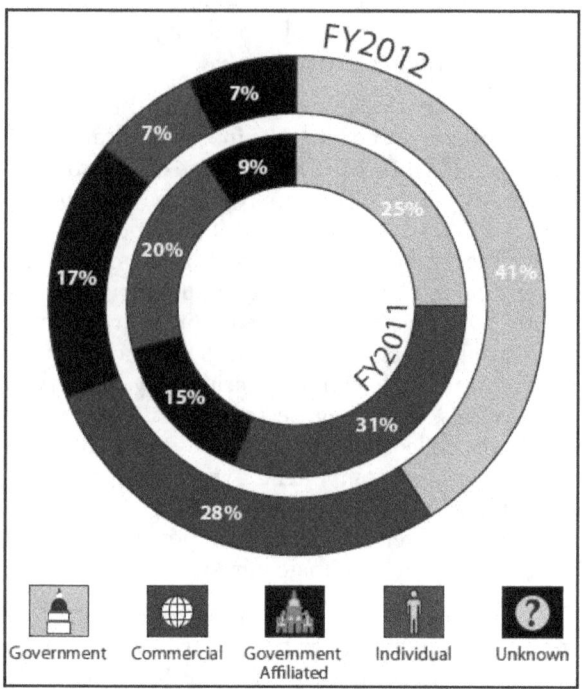

within a short period. For example, during a three-month period, three different East Asia and the Pacific firms made requests for the same electronics technology in comparable quantities.

Analyst Comment: As in FY11, given that many of these commercial entities requested specific technologies in very similar numbers and over short periods, they were probably acting on behalf of the same end user(s), ones that wished to avoid export-control complications should their identities or activities become known. (Confidence Level: Moderate)

As in FY11, a significant portion of the industry reporting linked to commercial entities consisted of requests to establish business relationships with cleared contractors. The number of these reports nearly tripled in FY12. Such reports were generally split between offers to act as the cleared contractor's regional distributor and as supplier of components to be incorporated into cleared contractor systems. One of the more common approaches involved East Asia and the Pacific companies offering to supply the cleared contractor with certain manufactured components, regardless of their applicability to the business activities of the contractor.

Analyst Comment: DSS assessed most of these offers as constituting legitimate business activity. However, the cleared contractors often produced technologies other East Asia and the Pacific entities were attempting to acquire. Acceptance of the offered East Asia and the Pacific components into U.S. defense system supply chains could lead to integration of poor-quality or altered materials that could degrade system effectiveness. Also, DSS cannot rule out that such approaches would lead to additional attempts at exploitation in the future. Therefore information regarding companies endeavoring to enter into such agreements is of intelligence value. (Confidence Level: Moderate)

The East Asia and the Pacific commercial entities that were active collectors in industry reporting, often working closely with government agencies, ranged from small, privately owned firms to major corporations. Some of the larger commercial entities have standing ties to government agencies that attempt to collect against U.S. technologies, according to reporting from another government agency (OGA). At least one has a dedicated and sophisticated technology collection directorate that works with domestic intelligence services to target foreign technologies.

Analyst Comment: The preeminence of commercial entities involved in suspicious activity from some East Asia and the Pacific countries likely is due to commercial sector entities acquiring technology for resale; commercial entities acting as front companies for prohibited end users, both in and outside the region; and governments using commercial entities and public tenders to cloak their involvement in targeting U.S. technologies. For those countries that enjoy favorable military and economic relationships with the United States, the desire to maintain those relationships probably partially explains government-affiliated entities accounting for relatively few suspicious contacts. (Confidence Level: Moderate)

After commercial, the next most common affiliations were government-affiliated (17 percent) and individual (seven percent). The majority of the reports in these two categories represented attempts to obtain postdoctoral or research positions with cleared contractors. As in FY11, most of these reports came from cleared contractors associated with U.S. universities. A large number of the requests came from professors or researchers at

East Asia and the Pacific universities or institutes, thereby placing them in the government-affiliated category; if DSS assessed the researchers as no longer attending/working at such a university, it placed the reports in the individual category. The remaining reports in the individual category represented instances in which DSS was unable to connect persons to any other entity.

Evaluation of the requests found that almost none of the researchers had any apparent connection to intelligence and security services. However, some reporting has noted certain East Asia and the Pacific students and academics exploiting access to sensitive or restricted technologies to support parallel research and development (R&D) efforts in their home countries. It is also noteworthy that even unaffiliated individuals' requests often mirrored those of East Asia and the Pacific commercial entities.

Analyst Comment: While it is very likely that most of these requests to obtain academic or research positions with cleared contractors were legitimate, DSS cannot rule out that some East Asia and the Pacific academics would ultimately take advantage of their placements to further national R&D goals. (Confidence Level: Moderate)

METHODS OF OPERATION

For the second year in a row, East Asia and the Pacific actors relied on SNA more than any other MO to target sensitive or classified information and technology resident in cleared industry, based on industry reporting. From FY11 to FY12, the number of East Asia and the Pacific-connected SNA cases increased over 240 percent and became even more prominent in the statistics, increasing from 23 to 42 percent of the total. DSS ascribed a large portion of East Asia and the Pacific SNA incidents to government entities.

East Asia and the Pacific cyber actors have varying levels of technical sophistication, but they are consistent in their malware delivery method. The reports DSS collected from cleared industry and government partners in FY12 overwhelmingly pointed to spear phishing as the number one methodology. This was the case in FY11 as well.

Analyst Comment: Spear phishing is likely the preferred tactic for any cyber actor to deliver malware because of its low cost-to-success ratio. Without much effort, actors can almost certainly

Figure 7: Methods of Operation

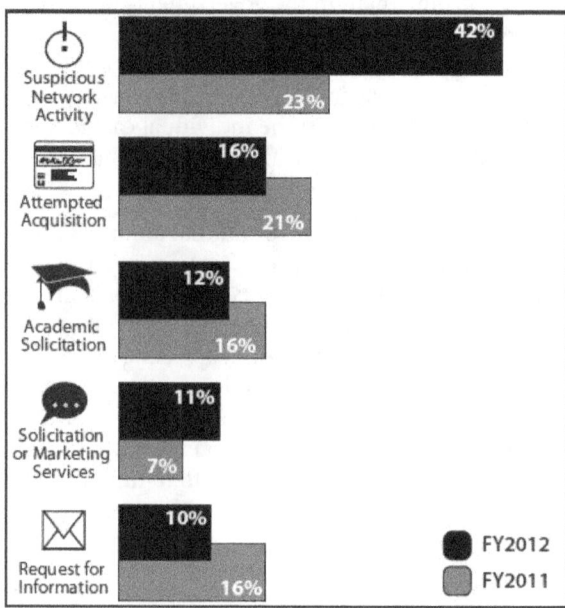

Method	FY2012	FY2011
Suspicious Network Activity	42%	23%
Attempted Acquisition	16%	21%
Academic Solicitation	12%	16%
Solicitation or Marketing Services	11%	7%
Request for Information	10%	16%

Figure illustrates the top five most reported methods of operation in FY12 compared to the same methods for FY11. None of the other methods totaled more than six percent of reporting.

distribute malware to an unlimited number of targets, which likely increases the chance of success. (Confidence Level: High)

AAT was the second most commonly reported MO, at 16 percent; RFI was fifth at ten percent. East Asia and the Pacific entities used both these MOs in very similar fashion. Typically, commercial entities contacted cleared contractors via email requesting sensitive components or information, often making no mention of the intended end user or use. DSS ultimately connected many requests that initially appeared relatively benign to East Asia and the Pacific militaries. As in FY11, a number of requests from East Asia and the Pacific commercial companies came through entities in third countries.

Analyst Comment: Most East Asia and the Pacific commercial agents very likely view AATs and RFIs as straightforward means of attempting to acquire information or technology they consider necessary. Emailing such requests poses little risk, while even an occasional success can provide a substantial reward. (Confidence Level: Moderate)

Academic solicitations from East Asia and the Pacific entities increased in number by 42 percent, but dropped from 16 percent of total industry reporting in FY11 to 12 percent in FY12. Cleared

contractors associated with U.S. universities reported the largest number of such solicitations. East Asia and the Pacific nationals sent the great majority of these requests via emails that included an introductory note as well as an attached résumé. Evaluation of these reports revealed that they originated from a broad range of universities, and that the same students often applied to multiple cleared contractors.

Analyst Comment: It is very likely that most of these instances of students and postdoctoral researchers from East Asia and the Pacific attempting to obtain positions at cleared contractor facilities were legitimate academic inquiries, and that the increase in numbers resulted from improved cleared contractor reporting. (Confidence Level: High)

DSS assesses that many East Asia and the Pacific students and academics in the United States probably pose a counterintelligence and technology transfer threat to cleared industry. While available information does not point to a direct connection between most, if any, academics and home-country intelligence services, such individuals and their sponsoring institutions likely view placement in U.S. facilities as supporting current R&D goals, some of which have military applications. Such placement opportunities are abundant in the United States, and East Asia and the Pacific students will almost certainly continue to seek them. (Confidence Level: High)

Solicitation or marketing accounted for 11 percent of the FY12 East Asia and the Pacific-connected industry reporting total, up from seven percent in FY11, and the number of such cases nearly tripled. It was practiced almost exclusively by commercial entities. In most incident reports citing this MO, an East Asia and the Pacific company offered to act as the cleared contractor's agent or distributor in the region or conveyed its desire to be the cleared contractor's main source of certain manufactured components.

Analyst Comment: There is an even chance that these proposals to form business partnerships were legitimate. However, if cleared contractors entered into such agreements, as a condition of the deal the East Asia and the Pacific business entity would likely request an exchange of personnel or even access to sensitive or classified U.S. information and technology, either of which could result in unauthorized access. (Confidence Level: Moderate)

Collectors from the region directed some solicitation or marketing inquiries toward cleared contractors producing LO&S or electronics, both technology areas in which East Asia and the Pacific collection efforts showed significant interest in FY12.

While the foreign visit MO accounted for only five percent of reported cases, the number of such cases increased by 43 percent from FY11. In one instance, a cleared contractor hosted a meeting with two government representatives and a commercial representative to discuss particular LO&S devices. However, the originally designated commercial representative did not attend and was replaced by an alternate. OGA reporting indicated the replacement employee, or a person of the same name, had served as an intelligence officer (IO) as recently as January 2011.

Analyst Comment: Last-minute changing of a delegation's roster is typical intelligence tradecraft for inserting an IO into a visiting delegation. DSS assesses there is an even chance the change in this case was intentional, to insert an IO into the delegation so as to use his experience to exploit the foreign visit in support of national collection priorities. (Confidence Level: Moderate)

TARGETED TECHNOLOGIES

The Intelligence Community (IC) has assessed that the long-term goals of regimes within East Asia and the Pacific include building the capability to field technologically advanced military forces. The IC has further assessed that doing so will require acquisition of a wide range of advanced systems, and/or development of indigenous capabilities to reduce dependence on Western countries. Acquisition of U.S. technologies may be accomplished either illicitly or in above-board fashion.

In FY12, East Asia and the Pacific entities targeted technologies in almost all sections of the MCTL. Combined, the four most commonly targeted areas accounted for over a third of the total attempts attributed to East Asia and the Pacific collectors: electronics (ten percent), IS (nine percent), LO&S (eight percent), and aeronautics systems (eight percent). The year-over-year jump by electronics to being the most commonly targeted technology area, more than doubling the number of reported incidents, marks a significant change from FY11, when electronics was the third most commonly targeted technology sector.

As noted in the FY11 version of this publication, East Asia and the Pacific collectors have demonstrated a strong interest in obtaining export-controlled U.S. rad-hard circuitry for several years. A substantial number of East Asia and the Pacific entities' requests for electronics technology targeted rad-hard integrated circuits. These circuits have applications in nuclear weapons, aerospace vehicles, ballistic missiles, and other electronics used in environments subject to radiation.

A number of East Asia and the Pacific countries have or intend to develop space programs and therefore have a perceived need for rad-hard, space-qualified circuitry. The lion's share of these requests noted specific circuits in precise quantities, but listed little to no end-use or end-user information.

Analyst Comment: The large number of requests that disparate East Asia and the Pacific commercial entities made for specific quantities and types of rad-hard circuits very likely denoted an ongoing need by customers that do not wish their identities disclosed. DSS has also noted similar efforts made through entities in other countries and regions. DSS assesses that if East Asia and the Pacific-connected collectors successfully acquired the requested materials, they would probably be used in their countries' space program(s). (Confidence Level: Moderate)

FY12 East Asia and the Pacific-connected requests for IS technology increased by nearly 20 percent in number of reported cases when compared to FY11, demonstrating continued interest in disparate systems, even though the sector's percentage of all industry reporting declined from 13 to nine percent.

The majority of IS-related attempts sought command, control, communications, computers, and intelligence (i.e., C4I) systems or, most commonly, modeling and simulation (M&S) systems. M&S and analysis software is often used in space systems, but East Asia and the Pacific collectors also sought these technologies for simulation centers and future systems research and development. For example, a representative of an East Asia and the Pacific government contacted a cleared contractor to discuss M&S products, voicing particular interest in software used in unmanned ground and air systems.

One frequently requested M&S software application facilitates satellite mission design, pre-mission analysis, orbit determination, collision avoidance, maneuver analysis, and formation flying. In multiple instances, East Asia and the Pacific entities requested M&S tools that use elaborate, distinct data and provide highly accurate two- and three-dimensional representations. While such software is often used for space systems, it can also be used for aircraft, missiles, ships, ground vehicles, and other moving objects. OGA analysis indicates that East Asia and the Pacific regimes seek simulation applications to enhance the warfighting capabilities of their armed forces. Further industry reporting over the past several years has corroborated East Asia and the Pacific interest in relevant simulation applications. East Asia and the Pacific collectors submitted a number of requests for such technology via webcard.

Analyst Comment: These requests very likely reflect East Asia and the Pacific regimes' continued perceived requirement to fill M&S needs, for both space-related and military applications. Language issues associated with some webcard forms forced DSS to assign some collection attempts to the individual affiliation, even though there is an even chance they originated with commercial entities. (Confidence Level: Moderate)

The number of reported East Asia and the Pacific-connected inquiries regarding technologies in the LO&S category increased by 44 percent over last year. Most attempts sought sensor technology that could address intelligence capability shortfalls, including inadequate intelligence, surveillance, and reconnaissance assets. Industry reports showed that East Asia and the Pacific entities targeted airborne, seaborne, and ground-based sensor systems in FY12. Shortwave infrared (SWIR) optical systems continued to be among the most commonly targeted items. SWIR optics are particularly useful for measurements imaging for a variety of civilian and military purposes, ranging from agricultural sorting and fault detection to hyperspectral imaging and handheld battlefield targeting.

Analyst Comment: The East Asia and the Pacific collectors probably sought to acquire sensor technology to enhance early warning capabilities. East Asia and the Pacific regimes likely intend to use any advanced radar and sonar systems gained from cleared contractors to mitigate the perceived threat from hostile neighbors. (Confidence Level: Moderate)

Notably, the MO for a significant portion of requests for LO&S changed from AAT or RFI in FY11 data to solicitation or marketing. Numerous East Asia and the Pacific companies contacted cleared contractors desiring to act as distributors for LO&S systems; to provide the cleared contractors with manufacturing facilities in the region; or to integrate their own technology, often optical, into the contractor's LO&S systems.

Analyst Comment: There is an even chance that the shift in MO reflected in industry reporting signifies East Asia and the Pacific collectors' efforts to improve access to sensitive U.S. LO&S-related information or technology, in response to a perceived lack of success of past AATs. Any relationship that offers access to cleared contractor systems or technology could lead to future opportunities for abuse, degradation, or exploitation of critical systems. (Confidence Level: Moderate)

Aeronautics systems technologies constituted the fourth most commonly targeted technology sector in industry reporting, accounting for eight percent of all reported attempts in FY12. Within the category, collection efforts focused on unmanned aerial vehicles (UAVs), including micro-air vehicles. In one instance, two East Asia and the Pacific-connected individuals questioned a cleared contractor employee at an annual armed service-related convention about micro-air vehicles. One of the individuals represented a government-affiliated institution, the other represented a commercial entity.

OGA analysis has assessed that East Asia and the Pacific regimes seek to improve their UAV capabilities with both medium- and high-altitude long-endurance systems. Successful collection against more capable U.S. systems would allow East Asia and the Pacific producers to provide protection to their countries at a lower cost and to undercut U.S. producers in the world market as well. Although the preponderance of aeronautic systems requests in FY12 involved UAV technology, East Asia and the Pacific entities also targeted manned fixed-wing and rotary-wing aircraft.

East Asia and the Pacific regimes seek to build ballistic surface-to-surface missiles with ranges sufficient to reach those other countries perceived as most likely opponents. Conversely, East Asia and the Pacific officials have stated their intent to upgrade surface-to-air missile systems to

Figure 8: Targeted Technologies

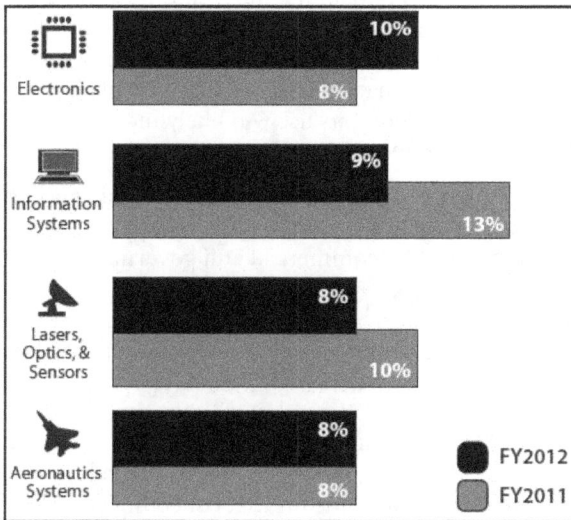

	FY2012	FY2011
Electronics	10%	8%
Information Systems	9%	13%
Lasers, Optics, & Sensors	8%	10%
Aeronautics Systems	8%	8%

Figure illustrates the top four most reported technology categories in FY12 compared to the same categories for FY11. None of the other categories totaled more than four percent of reporting.

counter the perceived ballistic missile threat from prospective opponents. See the special focus area of this publication for more discussion of targeters of missile technology.

When entities attempt to obtain illegal or unauthorized access to sensitive or classified information and technology resident in the cleared industrial base by attacking cleared industry networks, it is often difficult to ascertain which information or technologies they target. Department of Defense information on a cleared contractor network is often entwined with other company information, with no clear delineation between the two, except possibly for file names. Because the data is rarely segregated, it becomes very difficult to discern the target of the cyber actor. Additionally, while the cyber actor may target an individual for the information that resides on his or her local system, the actor may also make use of that system as a network entry point in pursuit of network resources. All these challenges make it exceedingly difficult in cyber cases to determine the specific technology targeted except during or after a compromise for which the data exfiltrated is known.

In 75 percent of the FY12 East Asia and the Pacific cyber cases reported to DSS from cleared industry and other government partners, the targeted technology was unknown. In those cyber cases

in which DSS did positively identify a targeted technology, aeronautics systems accounted for the most incidents, in seven percent of reports, followed by IS and marine systems at six and four percent, respectively.

Analyst Comment: The disparity in targeted technology between cyber and non-cyber cases may be due to collector affiliation. Non-cyber East Asia and the Pacific targeting of cleared industry typically involved a private entity not working directly for a government, even though a government may have been the intended ultimate end user of the technology. An example is the many offers to function as a reseller or manufacturer of SWIR-related technology within the region. Most of these requests are probably not made at the behest of the government; rather, private commercial entities likely make them while working for profit, with no direction from the government. (Confidence Level: Low)

Conversely, cyber reporting to DSS may reflect entities working directly for, or on contract to, a government to target cleared industry in cyberspace. In these cases, it is likely that East Asia and the Pacific collection entities actively target companies that align with governmental collection requirements, such as for aeronautics, and search for information resident on cleared industry networks to satisfy those requirements. (Confidence Level: Low)

Analysis of industry reporting detailing East Asia and the Pacific academics soliciting for research and postdoctoral positions at cleared contractors revealed that applicants came from a considerable range of educational and professional backgrounds. Frequently encountered elements included work or study in electrical engineering, computer science, materials science, and network security.

Analyst Comment: The backgrounds of many of the students applying for graduate and research positions are significant, in that they often show parallels with areas of interest at East Asia and the Pacific research and educational institutes, some of which maintain associations with national militaries. Student solicitations to study or work in such fields are of particular interest; DSS assesses that while most of these students have no connection to East Asia and the Pacific military or intelligence services, conducting research in the

United States very likely equips them to contribute to R&D in similar fields upon their return home. (Confidence Level: Moderate)

OUTLOOK

As projected in the FY11 Trends report, industry reporting originating from East Asia and the Pacific increased significantly during FY12. Notwithstanding recent political changes in the region, DSS assesses that East Asia and the Pacific entities will almost certainly continue to aggressively attempt to collect U.S. information and technology. DSS further assesses that East Asia and the Pacific will almost certainly remain the most prolific collector region in FY13 industry reporting, very likely continuing to increase in both number of reported cases and share of the total. (Confidence Level: High)

The increased reporting of East Asia and the Pacific efforts to obtain sensitive or classified U.S. information and technology likely demonstrates greater awareness of the threat on the part of cleared industry as well as the level to which East Asia and the Pacific R&D efforts continue to rely on foreign, particularly U.S., technologies. Whether the requesting entities are witting or unwitting participants, DSS assesses that much of any information or technology that East Asia and the Pacific-connected collectors acquired would ultimately be used to support the development of new defense systems or upgrade older ones. (Confidence Level: High)

In recent history, East Asia and the Pacific cyber actors have consistently accounted for the largest portion of attributed industry-reported SNA. Given the relatively poor state of network security across cleared industry, adversaries can make good use of even low-level capabilities, accomplishing a great deal with relatively little effort. It is almost certain that SNA originating from East Asia and the Pacific actors will continue, and that East Asia and the Pacific cyber actors will constitute the number one adversary for cleared industry in cyber space in the future. (Confidence Level: High)

East Asia and the Pacific cyber actors very likely consider that their continued success validates their cyber reconnaissance competencies, which probably allow them to design even more efficient targeting plans for the future, employing the full range of collection techniques. (Confidence Level: High)

Spear phishing emails containing malicious files or suspicious links remain the most efficient malware delivery mechanism, and will almost certainly remain the primary method for targeting cleared contractor information in FY13. Because of spear phishing's efficiency, its use will likely increase in FY13. (Confidence Level: High)

In addition to SNA, East Asia and the Pacific entities will almost certainly continue to employ overt requests made by commercial and government-affiliated collectors. RFI and AAT requests are seemingly innocent and often do not appear to have any connection to the military or R&D facilities, so they offer a low-risk, high-reward approach that collecting entities can employ without significant risk to relations with the United States. As noted in the FY11 Trends, the sheer continuing volume of requests probably means that collecting entities have achieved some success in using the AAT, RFI, and academic solicitation MOs. Foreign visits almost certainly provide excellent opportunities to attempt to collect against U.S. technologies. AAT, academic solicitation, solicitation or marketing, RFI, and foreign visit are likely to remain commonly employed methodologies for East Asia and the Pacific collectors. (Confidence Level: High)

The significant increase in reported solicitation or marketing collection attempts probably means that East Asia and the Pacific collectors are expanding their tradecraft. While many of the requests are doubtless legitimate, the increase in the use of this MO is noteworthy, especially as applicants directed many requests to cleared contractors producing technologies that East Asia and the Pacific entities previously or currently targeted. These requests could represent attempts by another means to gain access to the information or technology in question, or to access the U.S. supply chain. (Confidence Level: High)

Aside from the noted increase in the targeting of electronics systems technologies, which was very likely space-related, East Asia and the Pacific collection efforts did not appear to be focused in any one area in FY12. East Asia and the Pacific collectors again targeted a very wide range of U.S. technologies. As noted in previous editions of this publication, this targeting of such disparate technologies emphasizes the scope of modernization and R&D efforts going on within this region. Whatever alterations in defense policy emphasis may result from recent political changes within the region, long-term goals of achieving technologically advanced militaries will almost certainly mean the continuance of collection efforts aimed at a wide range of advanced systems from across the U.S. cleared industrial base. (Confidence Level: High)

For East Asia and the Pacific cyber actors, marine systems, IS, and aeronautics systems technologies will likely remain top collection priorities. These technologies are top targeted technologies in industry reports regarding SNA. It is likely that successful collection of the technologies would not only advance indigenous military modernization efforts but also lead to effective countermeasures to U.S. military technology. (Confidence Level: High)

CASE STUDY

The following case study illustrates the diverse and persistent approach typical of East Asia and the Pacific entities. Both commercial and government collectors targeted the IS technology in question, using both the AAT and foreign visit MOs.

In September 2011, a representative of an East Asia and the Pacific commercial entity emailed a cleared contractor requesting two different types of solid-state switches. He stated that the switches were intended for an indigenous radio system under development whose capabilities would be similar to a U.S.-produced system.

The representative also stated that his company was a contractor that does business with an agency of his country's government, and that the agency was looking for such switches for a related application. Several months earlier, a representative of the government agency had emailed a different cleared contractor seeking cooperation on developing an indigenous radio system like the U.S. product.

According to OGA reporting, the country in question has been interested in the U.S. system since 2004. The OGA assessment found that in 2009 the country funded a program to develop a system similar to the U.S. product program, to be completed in 2012. The assessment further determined that the foreign government agency frequently targets developing or existing U.S. technologies for collection and reverse-engineering, and judged that East Asia and the Pacific entities would probably attempt to reverse-engineer any related U.S. radio technologies it obtained and incorporate them into the indigenous system.

According to East Asia and the Pacific commercial representatives, the indigenous system will compete directly with the U.S. system, and the country in question began marketing components of its system as early as November 2009. An OGA assessed that the country in question would probably market its system as indigenous technology and continue to attempt to sell it to third countries.

Analyst Comment: There is an even chance that East Asia and the Pacific entities are interested in the U.S. radio system's technology because of an unmet milestone in the East Asia and the Pacific indigenous program, enhanced by the pressure of the original 2012 deadline. East Asia and the Pacific collectors likely remain interested in developing a system similar to the U.S. system for domestic military use and for commercial sales to third countries. (Confidence Level: Moderate)

NEAR EAST

43%
increase in
reported cases
originating in the
NEAR EAST

93%
increase in
reported
**government-
affiliated**
cases

Commercial
cases increased by
25%

Reports of
**attempted
acquisition** increased by **65%**

11% decrease in the number
of **requests for
information**

180% increase
in reported collection
attempts targeting
electronics

168% increase
in reported cases
targeting
**materials &
processes**

OVERVIEW

Based on industry reporting to the Defense Security Service (DSS) from fiscal year 2012 (FY12), Near East entities continue to be among the most active at attempting to obtain illegal or unauthorized access to sensitive or classified information and technology resident in the U.S. cleared industrial base. With 16 percent of the year's ascribed reports, the Near East was second only to East Asia and the Pacific. Targeting of U.S. technology attributed to Near East entities has steadily increased in volume over the last several years, and reported attempts increased again by over 40 percent from FY11. Overall, Near East collectors' affiliations, methods of operation (MOs), and targeted technologies discussed below remained consistent with Near East suspicious activity reported a year ago.

Geographically, the Near East is a relatively small area; yet it contains great political, economic, and cultural variety. Partly as a result, it is also home to significant frictions, resulting in international hostility within the region. Some of these differences have ramifications outside the region, resulting in additional frictions with states from other regions as well as world organizations. The frictions and the various resulting pressures on national regimes provide a great deal of the motivation to ensure the adequacy, if not the superiority, of defense forces.

Numerous Near East regimes, in turn, use this as justification for collection attempts to obtain illegal or unauthorized access to sensitive or classified information and technology resident in the U.S. cleared industrial base. International financial strictures and current national economic difficulties provide additional pressures to maintain and improve defense forces while expending as few resources as possible, including by enabling cheaper indigenous manufacture. As a result, Near East regimes continue to target and attempt to acquire U.S. information and technology illicitly.

Those entities that can openly gain access to U.S. cleared industry attempt to exploit personal contact with U.S. cleared contractors, especially by leveraging delegation visits to cleared contractor facilities. Others with less access instead use complicated networks consisting of intermediaries, procurement agents, brokers, and front companies to place orders through subsidiaries and intermediaries, obscure end-user information on shipping receipts, and operate in free-trade areas to avoid export-control regulations, according to industry and Intelligence Community (IC) reporting.

In both scenarios, the actor attempting collection against U.S. information and technology is seldom identifiable as a government agent per se, but some connection to a government is often discernible. Thus, government affiliated was the most frequently encountered category in FY12 industry reporting on the Near East, at 47 percent, even higher than

FY11's 35 percent. Based on industry reporting, the next most likely affiliation for a Near East collector was commercial at 28 percent, down slightly from 31 percent last year. This component represents Near East regimes' use of commercial entities as vehicles to pursue desired technologies. In FY12, individual collectors accounted for 11 percent of the total, down from 17 percent in FY11, and government collectors again accounted for around 10 percent.

DSS predicted in last year's edition of this publication that Near East-connected academic solicitations would continue at a high level in FY12, and in fact the number of such reported contacts increased by almost 80 percent over FY11. At 38 percent of the total, more Near East-connected reports fell into this method of operation (MO) category than any other. Attempted acquisition of technology (AAT) was the next closest at 22 percent. Request for information (RFI), solicitation or marketing, foreign visit, and seeking employment followed, ranging from 12 to 4 percent, respectively.

Students dominated Near East-connected academic solicitations with quests to conduct postgraduate-level research at U.S. academic centers involved in sensitive or classified Department of Defense (DoD) research. Such students sought positions in research programs in mechanical and aerospace engineering, materials science, and electrical and computer engineering. Evidence continues to surface of links between Near East governments and some students' academic interest in and solicitations to particular U.S. research programs. These connections between students and institutions help to explain why DSS assigned nearly half of FY12 industry reports of suspicious contacts related to the Near East to the government-affiliated category.

Similar to previous years, Near East entities' interest in U.S. technology encompassed nearly every category of the Militarily Critical Technologies List (MCTL). The most commonly targeted technologies largely remained consistent, with FY11's top four categories (information systems [IS]; lasers, optics, and sensors [LO&S]; aeronautics; and space systems technologies) remaining in the top six, being joined by electronics technology and materials and processes technology. All six of these technologies were bunched within a fairly narrow range, accounting for from 14 to eight percent of total industry reporting apiece, signifying the considerable dispersal of collector interest.

Within the pan-MCTL collection efforts that Near East collectors conducted, they displayed particular interest in acquiring modeling and simulation (M&S) software; aircraft components; unmanned aerial vehicle (UAV) platforms; radar technologies; dual-use electronic components; technologies and armaments associated with missile-defense platforms; and electronic warfare (EW) and signals intelligence (SIGINT) equipment.

COLLECTOR AFFILIATIONS

The nearly half of Near East-connected industry reporting categorized as government-affiliated included contacts linked to prominent educational institutions, research centers, and governmental defense firms. Government-affiliated entities increased their portion from 35 to 47 percent of the overall reported Near East collection attempts in FY12. Cleared contractors reported that these entities routinely made collection attempts, sometimes aggressively pursued.

For instance, an employee of a Near East entity who had previously worked with a cleared contractor on a U.S. Army aviation program began submitting probing emails to former acquaintances employed

Figure 9: Collector Affiliations

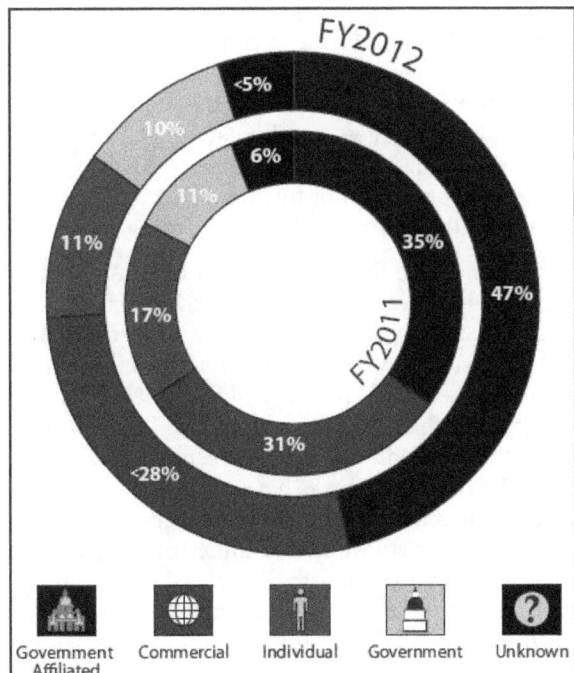

Government Affiliated — Commercial — Individual — Government — Unknown

with the cleared contractor. The individual initially asked follow-up questions regarding the product on which he had worked, but subsequently segued the conversation into detailed inquiries regarding the cleared contractor's other platforms.

Analyst Comment: DSS assesses that the individual was government-affiliated, and that his solicitations very likely constituted an attempt to exploit his existing relationship with the cleared contractor to gain unauthorized access to sensitive or classified U.S. information and technology pursuant to collection priorities of government-affiliated entities. (Confidence Level: Moderate)

Despite accounting for a slightly smaller percentage of total industry reporting in FY12 (28 percent) than in FY11 (31 percent), Near East commercial firms still accounted for the second largest portion of reported collection attempts. Emailing cleared contractors in attempts to procure information or technology characterized the majority of these commercial contacts. Some Near East firms contacted cleared contractors either on behalf of government-affiliated, government, or other commercial entities or in attempts to represent the cleared contractor in the region.

The degree to which commercial entities attempt to obfuscate their connections to and efforts on behalf of other potential collector entities varies within the region and over time as well. Connections with government agencies are most likely to be concealed. The varying degree of willingness to be identified with government procurement efforts can mirror overall relations between the home country and the United States. Another factor is whether the goods being sought are intended for use in the collector's country or for further distribution to third countries.

For instance, the further transfer of U.S. military technology to certain foreign countries requires prior U.S. approval. Some Near East countries have gained this approval on certain occasions for certain technologies; but there is some limited IC reporting to the effect that their conformance to U.S.-imposed limitations beyond that has not been consistent. Some Near East countries have relations, and have recently increased those relations, with countries in other regions to which the United States seeks to limit technology transfer. Near East companies that are willing to illicitly act as purchasing agents for these third countries even as they legitimately

acquire U.S. technologies for incorporation into their own country's defenses cause proliferation problems.

Conversely, some Near East countries continue to effectively divert prohibited U.S. technology to themselves through third countries in other regions with relaxed export-control laws and by exploiting established trade agreements between the United States and other countries. Front companies operating in these countries rarely attempt to contact U.S. companies directly, but rather rely on foreign entities to contact U.S. companies. Industry reporting to DSS continues to validate this pattern, showing requests for export-controlled technology originating from commercial companies that are associated with Near East countries but are located in almost all the other regions of the globe.

Other Near East collectors seek to evade trade limitations by requesting to purchase dual-use items, claiming civilian end-use applications.

Analyst Comment: Many of these Near East-linked requests for U.S. technology that industry reported in FY12 likely contained falsified end-user information. Near East commercial entities were probably not the intended end users of the requested technologies with military applications; instead, the true end users probably used these entities to mask their own identities and the projected military purposes. (Confidence Level: Moderate)

In one case of AAT by a commercial entity during FY12, a Near East national sought technical specifications and a product catalog for a wide-band tuner from a cleared contractor. The individual claimed to represent a Near East communications company and stated the intended use as PhD research. According to the cleared contractor, the wide-band tuner has military and intelligence applications.

Analyst Comment: The individual was likely employed by the communications company. The request for a wide-band tuner was probably not intended to fulfill research needs as claimed, but to meet the Near East government's SIGINT requirements. (Confidence Level: Moderate)

According to IC assessments, some Near East commercial firms cooperate directly with national intelligence services. Industry reporting of related incidents primarily involved delegation visits to cleared contractor facilities.

Analyst Comment: FY12 industry reporting portrayed such delegations as typically including Near East individuals with a record of soliciting cleared contractor employees for sensitive or classified U.S. information and technology. DSS assesses that such companies' prominence in reports of attempted collection activities during facility visits likely reflects cooperation with intelligence services. (Confidence Level: Moderate)

Individual collectors represented 11 percent of suspicious contacts reported from industry, a decrease from 17 percent in last year's reporting. In these cases, individuals involved in suspicious contacts omitted or denied any affiliation with Near East government, government-affiliated, or commercial entities.

Reported contacts ascribed to governments comprised ten percent of overall Near East incidents, consistent with FY11 results. Some collection attempts involved government personnel contacting cleared contractors via email seeking information or technology; contacting a cleared contractor directly, seeking to initiate collaboration; or participating in delegation visits to cleared facilities.

METHODS OF OPERATION

The most commonly used MO attributed to Near East entities continued to be academic solicitation, accounting for nearly 40 percent of industry reporting, up from 31 percent in FY11. Near East students typically sought postgraduate positions, thesis assistance, reviews of draft scientific publications, and access to U.S. research papers.

Based on industry reporting, DSS assesses that some of these Near East students seeking entry into postgraduate research programs at U.S. universities engaged in sensitive or classified DoD research did so under full scholarships from their governments, even though the governments in question might deny this. These funded scholarships tend to go to a very few carefully screened and selected students who seek to enroll in foreign graduate schools outside of their home country.

Analyst Comment: Near East governments probably provide these scholarships contingent upon these students returning to their home

countries and joining defense-related research and development programs to which they can transfer their acquired technical know-how. (Confidence Level: Moderate)

During FY12, instances of AAT attributed to Near East entities increased by nearly 65 percent over the previous year, and the category's proportion of the total increased from 19 to 22 percent. Near East entities sought via email to purchase U.S. technologies, including to download sensitive U.S. software programs. The example below illustrates the dangers these AATs can represent.

During FY12, an engineering manager for a Near East natural gas-fired power plant emailed a cleared contractor seeking a quote for valves that are export-controlled. Multiple European governments denied purchase and export requests for the valves.

Analyst Comment: The Near East entities in question likely had existing requirements for valves and were seeking alternative avenues to obtain them. The European officials probably denied the purchase and export requests for valves because of concerns they would be diverted to prohibited programs. (Confidence Level: Moderate)

Figure 10: Methods of Operation

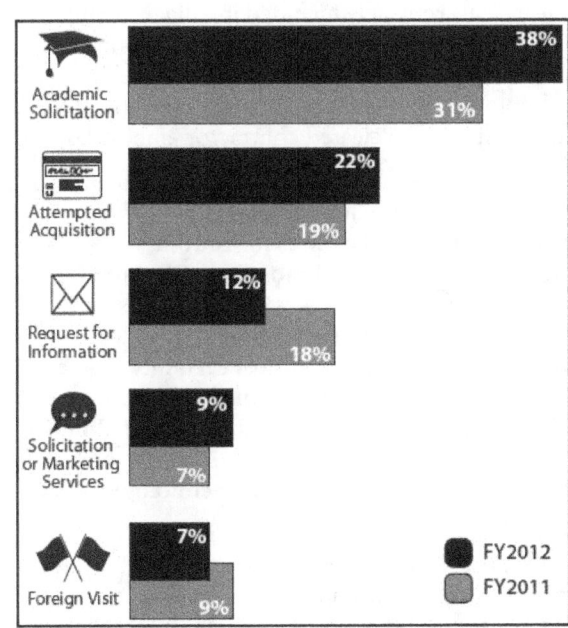

Figure illustrates the top five most reported methods of operation in FY12 compared to the same methods for FY11. None of the other methods totaled more than five percent of reporting.

In FY12, Near East entities sent RFIs to cleared industry through the use of web cards and unsolicited emails. These requests generally consisted of collectors seeking technical specifications, pricing, data sheets, and access to U.S. academic research papers. The Near East commercial sector also used email when employing the solicitation or marketing MO, which increased from seven to nine percent of total industry reporting in FY12.

While soliciting cleared contractors to market cleared contractor-developed products overseas constituted the most common version of this MO, Near East firms attempted to pursue business opportunities in numerous other incidents as well. The purported opportunities for collaboration with the United States extended across multiple defense sectors, including radar technologies, ground and air platforms, and analytical consultancy.

While foreign visit was only the fifth most commonly cited MO in Near East-connected industry reporting and decreased slightly in its percentage of the FY12 total, it nonetheless increased in number of reports. Near East government and government-affiliated entities practiced this MO primarily by leveraging official visits to a cleared facility to facilitate attempts to illicitly gain access to sensitive or classified U.S. information and technology. In FY12, such delegations: routinely included intelligence officers (IOs); arrived at cleared facilities with unannounced visitors; took unauthorized photographs of U.S. technologies; and attempted to elicit sensitive information from cleared contractor employees.

Analyst Comment: U.S. foreign-trade agreements typically enable the foreign company to visit cleared contractor facilities for routine visits and program reviews. If more Near East firms successfully solicit business opportunities with cleared contractors, delegation visits to cleared contractor facilities will almost certainly increase. Such delegations would very likely continue attempting to leverage such visits so as to obtain illegal or unauthorized access to sensitive or classified U.S. information and technology. (Confidence Level: Moderate)

While Near East entities' attempted exploitation of relationships accounted for few reported instances, such attempts often augmented collection operations during foreign visits and official travel. Near East entities of various affiliations attempted to leverage any personal contact, whether occurring in the United States or in the region, to target U.S. information and technology. This exploitation method usually involved asking questions that went beyond the agreed-upon scope of conversations.

For instance, a cleared contractor employee visited the Near East in FY12 to meet with foreign government officials regarding one of his company's products; however, once the employee was there the foreign personnel persistently expressed interest in another of the cleared contractor's products not covered by the governing agreement. The cleared contractor employee therefore declined to answer and advised the foreign personnel to use official channels.

TARGETED TECHNOLOGIES

Similar to last year, Near East collection efforts spanned all categories of the MCTL. Based on industry reporting, Near East collectors most often sought technologies related to electronics; IS; LO&S; materials and processing; aeronautics; and space systems. Reported collection attempts often sought specific legacy defense systems or their components, and DSS often ascribed them to government entities, which sometimes denotes a collection priority.

Reported cases of Near East collectors' targeting of electronics increased by 180 percent from FY11 to FY12. The electronics sector doubled its share of the total to 14 percent, and rose from being the sixth most targeted technology to the first. Based on industry reporting, Near East interest in electronics included U.S. communications, EW, and SIGINT equipment, including cognitive radios, digital receivers and decoders, demodulators, signal processing components, and direction-finding antennas. Reporting by other government agencies (OGAs) indicates that Near Eastern regimes seek advanced EW equipment to counter or compete with western technological advantages, and that they continue to seek to purchase foreign EW equipment even as they simultaneously attempt to develop and improve indigenous versions.

Additionally, in FY12, Near East collectors sought a wide range of dual-use electronic components. They did so almost entirely through intermediaries located in third countries. Near East entities also expressed interest in U.S. academic research programs related to directed energy.

During FY12, Near East entities targeted IS nearly as often as electronics, in 13 percent of all requests. DSS linked the majority of these requests to a substantial number of entities seeking to download M&S software used for space, defense, and intelligence systems from a cleared contractor's website. To gain access to the software, Near East entities often falsified end-user information by providing U.S. cities and organizations, which obfuscated their country of origin. Additional IS technologies of interest included wireless networking, network security, and other M&S software programs.

In FY12 industry reporting data, LO&S technologies accounted for 10 percent of the total. Given the geopolitics of the region, numerous countries fear airborne and missile attacks from neighbors and others, but none possesses fool-proof defensive systems against them.

Analyst Comment: Deficiencies in national missile and rocket detection capabilities likely explain Near East collectors' emphasis on targeting U.S. radar technologies. DSS assesses that Near East entities will very likely continue to target radar technologies to attempt to maintain an advantage against likely adversaries. (Confidence Level: Moderate)

Near East entities targeting materials and processing technology also accounted for 10 percent of all requests, and the number of such reports almost doubled from FY11. These requests almost entirely came from students seeking entry into one particular U.S. university's materials science research program. Several students requesting admittance to this U.S. university stated that a government agency had provided them with scholarships; one applicant stated that he received additional funding from a quasi-governmental foundation. Financial assistance was accompanied by military service waivers to facilitate foreign study by exceptional students conducting research in science and technologies that had military applications.

Analyst Comment: Graduate research scholarship students likely apply to the cleared contractor programs because of the programs' military applications, and if admitted would almost certainly transfer any technological knowledge acquired back to their countries. (Confidence Level: Moderate)

Aeronautics systems technologies, especially UAV platforms and associated technologies, specifically surveillance and reconnaissance unmanned aerial systems (UAS), were subjected to targeting. Access to such U.S. technologies could improve indigenous Near East capabilities, enabling them to remain competitive in the global marketplace.

Near East commercial firms and government entities contacted cleared contractors on multiple occasions requesting detailed technical information on particular UAS programs. According to 2010 OGA information, Near East countries sought UAVs capable of providing real-time intelligence over border areas, including UAVs with a "higher level capability," such as that provided by U.S. systems.

Analyst Comment: Foreign customers' preference for more advanced U.S. UAS over Near East indigenous systems is probably provoking Near East-origin collection efforts. For some in the region, a determination to maintain a competitive economic advantage in UAV platforms likely prompts attempts to collect against comparable U.S. systems. (Confidence Level: Moderate)

Figure 11: Targeted Technologies

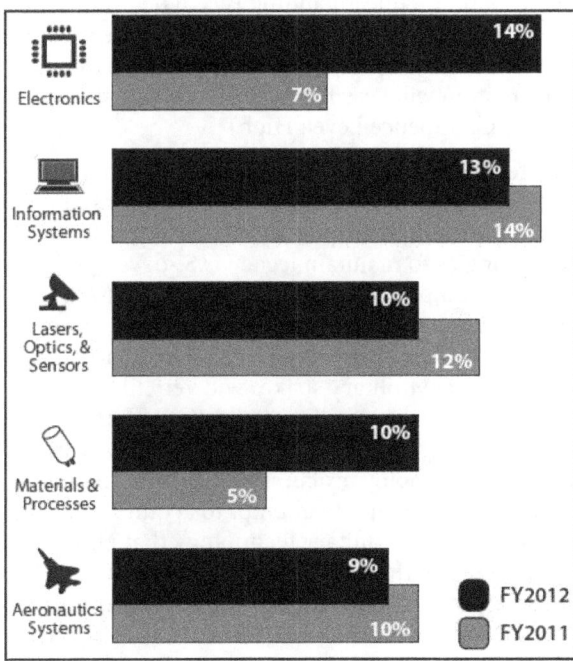

Figure illustrates the top five most reported technology categories in FY12 compared to the same categories for FY11. None of the other categories totaled more than eight percent of reporting.

In FY12, industry reporting documented Near East-connected collection attempts to obtain unauthorized information about armaments and energetic materials technologies related to missile defense platforms that Near East end users had previously purchased. For instance, an employee of a Near East firm made multiple attempts to procure technical specifications for a bomb release unit that fell under the International Traffic in Arms Regulations (i.e., ITAR).

Analyst Comment: Near East entities very likely seek to acquire specific conventional munitions based on perceptions of their usefulness should regional antagonists conduct strikes against each other and experience mutual retaliation. (Confidence Level: Moderate)

OUTLOOK

DSS assesses that Near East entities will likely remain aggressive in their attempts to collect U.S. information and technology. Owing to both general hostility from neighboring countries and specific regional concerns, Near East collectors almost certainly perceive a need to acquire the most effective technologies available to counter threats from short- and long-range rockets, weapons smuggling, and cyber and nuclear attacks. Accordingly, Near East entities will very likely continue to justify exploiting every opportunity to attempt to obtain illegal or unauthorized access to sensitive or classified information and technology resident in the U.S. cleared industrial base so as to enhance their countries' defensive measures. (Confidence Level: High)

International trade limitations and economic and budgetary difficulties will likely hamper Near East collectors' ability to illicitly acquire U.S. technology in the future, but may increase collection efforts. Collectors are least likely to allow these difficulties to interfere with attempts to obtain U.S. technology for programs, such as air defense systems, that regimes deem critical to national perseverance. (Confidence Level: Low)

The hierarchy of affiliations attributed to Near East collectors will likely remain unchanged during the next fiscal year. Financial limitations may cause the number of contacts by government-affiliated entities, in the form of academic solicitations specifically linked to proposed attendance at U.S. universities, to level off. But Near East students will likely continue to attempt to exploit the academic community by soliciting assistance on fundamental and developing research and attempting to acquire dual-use components under the guise of academic research. (Confidence Level: Low)

DSS assesses there is only an even chance that the Near East's commercial sector will become more prominent in industry reporting as procurement agents for domestic and third-country end users. However, some Near East countries acquiring U.S. technology legitimately could serve as conduits for transfer to countries in Europe and Eurasia and East Asia and the Pacific. (Confidence Level: Moderate)

The MOs used in Near East-connected collection attempts targeting cleared contractor facilities will likely remain consistent in the near future, with academic solicitations, then AATs, then unsolicited RFIs as most common. Academics have plausible reasons to request information and technology, often under the guise of scientific research. Near East collection efforts will probably continue to exploit the region's academic population, which generates a large percentage of reported suspicious contacts. (Confidence Level: Moderate)

Official Near East delegations will very likely continue to use visits to cleared facilities as a vehicle to target U.S. information and technology. Use of such tactics as soliciting information beyond the agreed-upon scope, including IOs in delegations, and using prohibited phones, recording devices, and cameras to attempt to collect unauthorized information will very likely persist during these visits. (Confidence Level: High)

The technologies that Near East collectors target will probably remain consistent as militaries in the region seek to acquire a wide variety of U.S. technologies to maintain legacy U.S.-developed military equipment as well as to carry out force modernization. Current international pressures and the perceived threat of inter-regional and intra-regional military strikes will very likely continue to drive Near East collection efforts. To this end, Near East collection efforts will very likely continue to employ procurement agents, brokers, and front companies to attempt to acquire U.S. technology. Any military technology that Near East commercial entities acquire will likely be diverted to government and/or military entities. (Confidence Level: Moderate)

When applicable, Near East entities will likely continue to attempt to procure dual-use items for chemical and nuclear programs under the guise of procurements for energy and gas industries. (Confidence Level: Moderate)

In an era of declining defense budgets, DSS assesses that collection attempts against technologies that Near East regimes previously acquired from the United States are likely to increase. Such targeting of restricted information on exported U.S. technologies, when successful, almost certainly allows Near East entities to minimize deficiencies in and otherwise strengthen indigenous capabilities. (Confidence Level: Moderate)

While Near East collectors tend to prioritize their efforts toward improving their countries' defensive measures, DSS assesses there is still an even chance that Near East entities will attempt to collect against certain technologies for purposes of enhancing economic competitive advantage. (Confidence Level: Moderate)

CASE STUDY

The following case provides insight into Near East use in illicit collection efforts of procurement networks consisting of front companies, intermediaries, and brokers located in various countries. It also illustrates the persistence characteristic of these efforts, as it involves repeated AATs of U.S. microwave electronics from the same cleared contractor.

In July 2012, a representative of a commercial firm in Europe and Eurasia contacted the cleared contractor requesting a minimum of 20 microwave electronic components. He did not identify an end user but stated the devices were for his clients.

That same month, a foreign sales representative of a firm in East Asia and the Pacific contacted the cleared contractor requesting a quote for 50 pieces of microwave electronics for an unidentified end user. According to an OGA, the firm is likely involved in the illegal sale of export-controlled U.S. technologies to its home country.

Also that month, a commercial manager of a Near East firm contacted the cleared contractor requesting a quote for 50 pieces of the same microwave electronic components. A 2011 OGA report found that this company appeared on a list of Near East companies that requested circuit boards for control systems from an identified U.S. business.

Also that month, an individual claiming to represent a consortium from East Asia and the Pacific contacted the cleared contractor seeking 20 pieces of microwave electronic components for a client in Europe and Eurasia. Open sources confirmed the individual was the point of contact for the consortium, and business listings for the company cataloged it as a Near East company with locations there and in East Asia and the Pacific. The company purportedly specializes in edible commodities. Notably, however, a 2010 OGA report identified this individual as the point of contact for the first Europe and Eurasia company mentioned at the beginning of this case study.

The next month, an individual representing a different company in East Asia and the Pacific contacted the cleared contractor requesting a quote for 50 of the same microwave electronic components for his 'customer.'

Analyst Comment: DSS assesses that these different requests for almost identical quantities of the same part number in close succession constituted suspicious contacts, and probably represented Near East procurement agents' efforts to illicitly acquire U.S. microwave components. Three of the five companies involved have ties to the same Near East regime. While one East Asia and the Pacific company illicitly procures components for end users in its home region, it cannot be discounted that the company is also involved in illicit procurement on behalf of countries in other regions as well. (Confidence Level: Moderate)

SOUTH AND CENTRAL ASIA

12 percent of FY12 reporting that originated in **SOUTH & CENTRAL ASIA**

That is a **67%** increase in the number of cases

239% increase in **government-affiliated** cases

Commercial cases increased by **27%**

341% increase in reports of South & Central Asia entities utilizing **academic solicitation**

62% increase in reports of attempted acquisition

106% increase in reports of seeking employment

196% increase in reported **materials & processes** targeting

148% increase in reported cases targeting **electronics**

OVERVIEW

Fiscal year 2012 (FY12) industry reporting showed that suspicious entities linked to South and Central Asia were again among the most prolific in attempts to obtain illegal or unauthorized access to sensitive or classified information and technology resident in the U.S. cleared industrial base. The number of industry reports the Defense Security Service (DSS) attributed to South and Central Asia entities in FY12 increased by nearly two-thirds over the previous year, although they continued to account for the same percentage of world totals. Based on industry reporting of suspicious incidents, South and Central Asia moved to third place on the list, displacing Europe and Eurasia.

This surge likely resulted from an escalatory "ratcheting effect" between rival regimes and collectors within the region. However, parsing the geopolitical dynamics of the region is complicated by the impact of other actors that, while external to the region, have a significant effect on the policies pursued within it, by their alignment with certain actors and against others. This has the effect of causing some regional actors to target U.S. technology in cooperation with external actors, and other regional actors to do so to attempt to counter the enhanced effect of such cooperation.

An additional factor is regional perceptions of U.S. foreign policy priorities. Shifting perceptions of the U.S. along with influence from states outside the region may lead some South and Central Asia states to reevaluate their willingness to rely on U.S. military technology. Despite any such changes that may occur regarding perceptions of the United States and its military technology, a significant threat of transfer of U.S. defense technology, whether acquired legally or illicitly, will likely continue.

Motivated by regional instability associated with neighboring Afghanistan, counterinsurgency efforts, and countering each others' strategic goals, multiple South and Central Asia states are attempting to maintain and modernize their militaries by purchasing new technologies or upgrading and replacing older systems. However, with national economies still performing poorly in general, the negative impact on defense budgets will potentially increase the illicit targeting of western, including U.S., technology.

South and Central Asia entities have a history of providing false or misleading information to attempt to acquire dual-use technology for restricted applications. We cannot dismiss the possibility that some of the seemingly legitimate South and Central Asia interest in cleared contractor products, technologies, and personnel represented attempts to acquire sensitive or classified information and technology for prohibited purposes or end users.

In FY11 commercial entities accounted for the largest proportion of South and Central Asia-connected reports from industry, at 47 percent, and the number of such cases increased in FY12 by over 25 percent. However, the number of cases ascribed to government-affiliated entities increased even more over the same period, more than tripling, so the latter category accounted for 37 percent of the FY12 total, just higher than the commercial segment at 36 percent. In contrast, in FY11 industry reporting, the government-affiliated category accounted for only 18 percent of the total. The numbers of reports ascribed to individual and government entities also increased in FY12, but both decreased their shares of the total, accounting for 17 and six percent, respectively.

Based on FY11 industry reporting, attempted acquisition of technology (AAT) was the most common method of operation (MO) for South and Central Asia collectors, and solicitation or marketing was the fifth most common; those MOs held the same positions in the FY12 listing, at 31 and eight percent, respectively. In between those two, last year's fourth most common MO, academic solicitation, increased its share from nine to 25 percent of the total and climbed to the second position. This pushed request for information (RFI) and seeking employment down to the third and fourth spots, with 18 and 12 percent of the total, respectively.

Looked at another way, among the five most common MOs represented in FY12 industry reporting related to South and Central Asia, while all increased in number of reported cases, three (AAT, RFI, and solicitation or marketing) declined in shares of the total, while two (academic solicitation and seeking employment) increased their shares of the total. Statistics for the latter two MOs often move in parallel with each other.

Instances of AAT, the most commonly encountered MO in industry reporting related to South and Central Asia, largely involved commercial entities acting as procurement agents seeking U.S. technology and military equipment, and who identified the military or some other government entity as the intended end user. This overt interest in U.S. technology was even more apparent in the volume of academic solicitations from South and Central Asia government-affiliated university entities to subject matter experts (SMEs) in U.S. industry.

South and Central Asia entities continued to display a wide-ranging interest in U.S. information and technology. Comparing FY11 to FY12 statistics regarding technologies targeted reveals a mixed and dynamic picture. Among last year's top eight targeted technologies, information systems (IS) and aeronautics experienced a decrease in both their number of reported cases and proportion of the total, and they fell from being the first and third-most targeted sectors to third and sixth. Two other sectors, lasers, optics, and sensors (LO&S) and positioning, navigation, and time, increased in number of reports but fell in their shares of the total, although they maintained their approximate positions in the rankings. Four other technology sectors, electronics, materials and processes, processing and manufacturing, and space systems technologies, increased both in number of cases and their shares of the total. Electronics jumped from being the fourth most targeted technology sector in FY11 industry reporting to the top position in FY12.

The wide-ranging interest that South and Central Asia entities showed in U.S. information and technology almost certainly reflects the numerous military modernization efforts that states in the region are conducting for which they perceive a need for western technologies. DSS judges that South and Central Asia collectors very likely intended for many of the targeted technologies to be used to upgrade older variants currently in use. Government efforts to establish domestic defense industries capable of providing weapons systems and technology on par with those of western companies will likely lead to attempts—characterized by varying degrees of discretion—to reverse-engineer any U.S. technologies acquired.

COLLECTOR AFFILIATIONS

DSS attributed over a third of industry-reported incidents with a South and Central Asia nexus to government-affiliated entities, marking a doubling of this affiliation's share of the total in one year. Seemingly legitimate commercial entities initiated the majority of requests that DSS categorized as government-affiliated.

Analyst Comment: DSS assigned AATs from commercial entities whose actions reflected official government inquiries to the government-affiliated category. DSS assessed that the majority of the

reported government-affiliated suspicious contacts probably constituted efforts to meet government requirements. (Confidence Level: High)

DSS attributed a number of government-affiliated requests, as well as some solicitations perpetrated by individuals, to students, professors, researchers, and engineers from state-funded universities, technology institutes, or research organizations. They made unsolicited contacts to fellow researchers at U.S. cleared contractors seeking technical information, employment, assistance with research, or the purchase of U.S. technology. Many students persistently pursued postdoctoral, internship, and research opportunities under technical SMEs employed at cleared facilities or at institutions of higher learning in the United States.

Graduate students, researchers, and faculty members at such government-affiliated South and Central Asia entities are also involved in military-related research. These institutions tend to provide technical education, offering multiple programs focused on disciplines such as engineering, chemistry, and physics. South and Central Asia government agencies responsible for military technology development have plans to increase their outreach to and collaboration with such research and educational institutions.

Commercial entities also accounted for over a third of industry-reported incidents with a South and Central Asia nexus, which represented a decrease from a 47 percent share of the total last year, although the number of cases increased by over 25 percent. From FY07 until this year, commercial entities had consistently accounted for roughly half of all reported South and Central Asia-originated collection attempts. Reported contacts from South and Central Asia commercial entities mainly consisted of RFIs or AATs directed toward business development and sales personnel in cleared industry.

Analyst Comment: DSS assessed that many of these inquiries originating in South and Central Asia probably constituted responses to official government requirements. (Confidence Level: Moderate)

South and Central Asia commercial entities identified government and military organizations as the intended end users in numerous requests. However, in many cases DSS could not verify the end user information provided, and in some instances could not find any information identifying the requestors. According to DSS analysis of

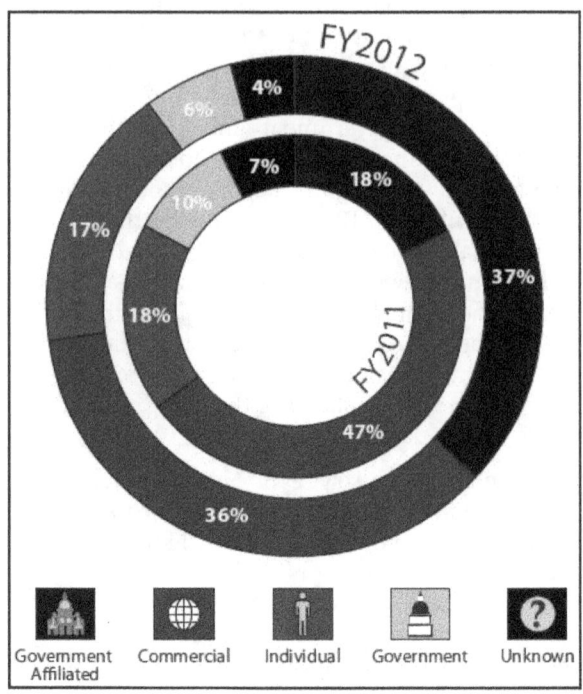

industry reporting, those commercial entities that did not identify an end user tend to have a history of making requests on behalf of South and Central Asia government organizations. Commercial entities sought a wide range of dual-use technologies across the various Militarily Critical Technologies List (MCTL) categories.

Analyst Comment: The lack of corresponding information on government requirements in the requests originating from commercial entities, particularly entities lacking identifying information, suggests the entities were attempting to procure technology on behalf of organizations that are on the Commerce Department's Entity List or have interests inimical to the United States. Further, DSS cannot rule out that these commercial entities were acting on behalf of unknown third countries from other regions. (Confidence Level: Low)

Individuals whom DSS assessed as acting on their own accord accounted for 17 percent of the South and Central Asia-related reports that cleared industry submitted during FY12. These individuals were primarily job seekers who applied for positions throughout cleared industry. While the threat these individuals posed was minimal, as they are ineligible for the sensitive positions for which they applied due to requirements for U.S. citizenship or security clearance, the sheer volume of those attempting to gain employment at research centers and cleared contractors engaging in high technology research and development (R&D) is noteworthy.

Analyst Comment: While the individuals' motivations for seeking employment in sensitive positions in cleared industry are unknown, DSS assesses there is an even chance these incidents were related to South and Central Asia governments' targeting efforts. (Confidence Level: Moderate)

However, some South and Central Asia governments do encourage their nationals residing in the United States with technical backgrounds to return their expertise to their home countries, presenting the risk that any sensitive information to which these individuals are exposed will be transferred.

Finally, government entities accounted for only six percent of industry-reported incidents with a South and Central Asia nexus. The majority of government-attributed contacts originated from scientists and engineers employed by agencies engaged in defense and/or space-related endeavors. During FY12, employees of such entities made numerous direct inquiries to cleared contractors involved in space and communications systems technologies. In general, South and Central Asia government contacts to cleared industry were overt, in almost all cases identifying both an end use and end user.

METHODS OF OPERATION

AATs and RFIs collectively gave rise to almost half of industry reports attributed to South and Central Asia-connected entities. Although these two MOs comprise separate categories, they are very similar in use. The incidents captured in these reports primarily involved email contacts to cleared contractor business development and sales employees. The vast majority of incidents categorized as AATs involved requests for specific components or platforms, while RFIs involved requests for pricing or technical specifications.

Analyst Comment: South and Central Asia procurement agents very likely view AATs and RFIs as legitimate means for successfully obtaining U.S. information and technology. DSS assesses it is very likely that many of these agents are genuine, but that a nefarious end user occasionally initiates a request. As a region, South and Central Asia generates a sufficient volume of inquiries that illicit contacts can be difficult to discern within the flow of legitimate requests. (Confidence Level: Moderate)

Academic solicitation stood out for the increase this MO experienced during FY12. The share that this category accounted for went from nine to 25 percent of the whole, and the category went from the fourth most commonly reported to the second. These solicitations primarily consisted of South and Central Asia nationals seeking research positions under high-profile researchers, scientists, and SMEs employed at cleared contractor components of academic institutions. A substantial fraction of the soliciting individuals had a connection to South and Central Asia state-funded universities, technology institutes, or research organizations, which boosted the government-affiliated category to the top of the affiliations list in FY12 industry reporting.

A substantial number of the recipients of these solicitations served as faculty members teaching technical courses at their academic institutions while at the same time conducting classified

Figure 13: Methods of Operation

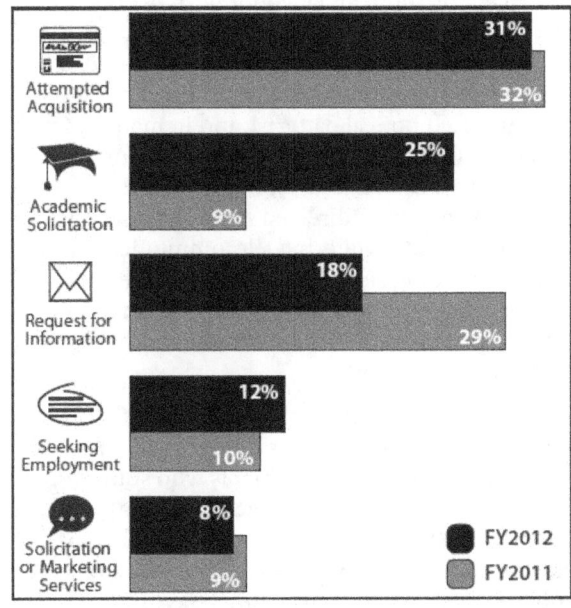

Figure illustrates the top five most reported methods of operation in FY12 compared to the same methods for FY11. None of the other methods totaled more than three percent of reporting.

research on behalf of a U.S. government customer. Many of the recipients have prominent academic status and much of their unclassified work appears in scientific or technical journals.

Analyst Comment: The involvement in classified U.S. government projects of professors and researchers receiving solicitations from South and Central Asia academics is not publicly available information. The South and Central Asia nationals in question were probably unaware which American researchers were involved in classified programs. However, it is likely that some of those making academic solicitations to cleared SMEs did so with the goal of gaining access to sensitive research efforts. (Confidence Level: Low)

We judge it is likely that South and Central Asia students and researchers are knowledgeable about American SMEs' unclassified work in their academic fields and that many of the academic solicitations reported represent legitimate efforts to study at American academic institutions that regularly admit foreign nationals. However, regardless of the legitimacy of the intent behind these academic solicitations, foreign nationals studying under or regularly interacting with cleared employees engaged in classified R&D almost certainly pose a threat. (Confidence Level: Moderate)

The seeking employment MO, which accounted for 12 percent of South and Central Asia-attributed reports from industry, involved attempts to apply for positions in cleared industry that required U.S. citizenship or a security clearance. These requests demonstrated no real patterns, and in many cases the entities' applications correlated with publicly available vacancy announcements. South and Central Asia entities directed some requests toward cleared contractors offering telecommunications and shipbuilding services. The prospective employees' backgrounds varied significantly, from engineers to information technology (IT) professionals.

Analyst Comment: As with the academic solicitation MO, we judge it is likely that some South and Central Asia nationals who sought employment at cleared contractors did so with the goal of gaining access to sensitive research. However, in many cases DSS judged that the applications cleared contractors received were unsolicited résumés that the applicants probably sent in bulk to multiple potential employers in

the United States. **Better awareness among cleared contractors concerning foreign individuals seeking employment likely contributed to the increased reporting of receipt of foreign résumés in FY12. (Confidence Level: Moderate)**

During FY12, cleared industry reported several incidents in which South and Central Asia entities sought to exploit existing or developing business relationships with cleared contractors to gain maximum access to sensitive processes, technologies, or products. While this behavior from foreign entities that purchase U.S. defense technology is not unusual, governmental determination to develop domestic military technology and systems contributes to the likelihood of reverse-engineering attempts on technology sold to South and Central Asia entities.

Analyst Comment: Cleared contractors are becoming increasingly involved in defense markets in South and Central Asia. This likely expands opportunities for South and Central Asia customers to exploit business relationships to maximize their access to U.S. technology. (Confidence Level: Moderate)

TARGETED TECHNOLOGIES

According to open sources, multiple South and Central Asia governments have launched long-term efforts to develop competitive domestic defense industries capable of meeting the breadth of their countries' military equipment and technology needs. In addition to these broad efforts, multiple South and Central Asia governments are currently undertaking a variety of high-profile, high-tech military R&D programs. These include aeronautical vehicle and missile R&D programs.

Many of the AAT incidents highlighting South and Central Asia entities' interest in specific U.S. systems and technology involved requests for small quantities of systems. Attempted procurement of small quantities of restricted technology points to an intent to reverse-engineer. DSS received reports during FY12 of suspected reverse-engineering of an electronics device legitimately sold to a South and Central Asia government agency. According to the report, during a conference at which the agency's engineers described a newly designed electronics device, several cleared contractor engineers recognized unique design features of a product they had previously sold to the agency.

Analyst Comment: The ambitious South and Central Asia goals combined with the history of reverse-engineering generate DSS concern over exploitation and reverse-engineering of any U.S. military technology that some South and Central Asia entities acquire. We judge that any military systems that these South and Central Asia entities acquire, legitimately or otherwise, are probably at high risk for reverse-engineering. (Confidence Level: Moderate)

In support of these ongoing government-funded programs, South and Central Asia entities continue to show interest in a diverse range of military and dual-use technologies that cleared industry produces, with no specific technology category standing out in industry reporting as a focus of interest. During FY12, the top eight MCTL categories collectively accounted for 75 percent of industry reports in which DSS could identify South and Central Asia interest in specific technologies. The top four (electronics; LO&S; IS; and materials and processes technologies) accounted for from 15 to ten percent apiece, and accounted for more than half the total.

Reported South and Central Asia-connected attempts to collect electronics technology more than doubled in FY12. In one example, industry reporting revealed many requests for the same quantity of certain radio components for unidentified end users. One such request originated from a South and Central Asia company that was previously identified as a front company used to procure electronic components for missile and nuclear programs. See the special focus area of this publication for a more detailed discussion of this issue.

Analyst Comment: The channeling of the requests for the radio components through the front company points to the intended end user being South and Central Asia nuclear or missile industries. Further, advancement of indigenous ballistic missile programs likely relies heavily on foreign materials and technology. South and Central Asia use of front companies in collection efforts has very likely been a successful means of gaining sensitive or classified dual-use U.S. information and technology from cleared industry. (Confidence Level: Moderate)

The LO&S technologies that South and Central Asia collectors targeted in FY12 included radar technologies and weapons sighting equipment. Radar technologies stood out due to the volume of requests from procurement agents. The requests appeared to correspond to public government requirements for airborne electronic warfare equipment.

Although the requestors did not specifically identify the radar technology sought, the cleared contractor involved presumed the targeted technology to be a particular airborne radar warning receiver (RWR) it produces. This RWR can detect radars associated with surface-to-air-missiles, airborne interceptors, and anti-aircraft weapons systems. A number of different South and Central Asia platforms, both ground-based and airborne fire-control systems, use RWRs.

Analyst Comment: Although the South and Central Asia-connected collectors may have made the RWR requests in response to a particular planned upgrade in an existing airborne system, there is an even chance the equipment would have been used to upgrade multiple platforms. The collecting entities may also be interested in procuring the cleared contractor RWR as an alternative to indigenous systems. (Confidence Level: Low)

Industry reporting shows that South and Central Asia entities also attempted, through U.S. middlemen, to acquire electron tubes manufactured for inclusion in their existing radar systems.

Figure 14: Targeted Technologies

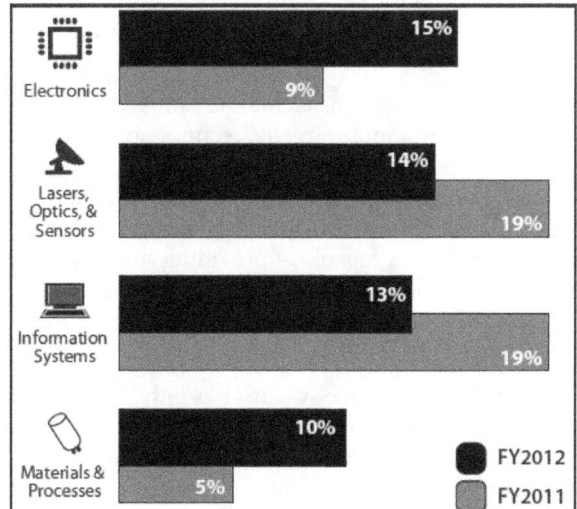

Figure illustrates the top four most reported technology categories in FY12 compared to the same categories for FY11. None of the other categories totaled more than eight percent of reporting.

Analyst Comment: The requested component could be used to upgrade older radar variants in use in South and Central Asia. South and Central Asia regimes are very likely attempting to update early warning systems and platforms to combat a perceived constant threat of air strikes. (Confidence Level: Moderate)

Many of the South and Central Asia-linked reports ascribed to the IS category involved individuals seeking employment in the IT industry. Many of the individuals were system or network engineers, while some were hardware and software designers and implementers. The IS-linked requests that focused on particular technologies sought integrated software systems, including for satellite and marine communications.

Analyst Comment: There is an even chance these communications systems would have been used in unmanned aerial vehicles (UAVs). (Confidence Level: Moderate)

Although the category of aeronautics systems technology fell to sixth position in FY12, South and Central Asia entities demonstrated continued interest in UAVs. Indigenous South and Central Asia unmanned aerial systems programs have experienced difficulties in developing advanced systems. South and Central Asia collectors have used both domestic and U.S.-based front companies in attempts to acquire more advanced systems.

Analyst Comment: South and Central Asia-connected collectors' continued targeting of cleared contractor UAVs almost certainly reflects efforts to support and expand force modernization plans and upgrades. (Confidence Level: High)

Although space systems technology ranked as only the seventh most targeted technology sector in FY12 industry reporting, it looms large in long-term planning for South and Central Asia regimes. Such plans include the development of low-cost satellites capable of providing space-based surveillance and command and control. Related interests include space security and counter-space technologies.

This emphasis on space systems has only just started to manifest itself in reporting from cleared industry regarding South and Central Asia. The number of South and Central Asia-connected cases relating to space systems technology more than doubled from FY11 to FY12. As an example, during FY12 a South and Central Asia government agency contacted multiple cleared contractors seeking space systems

and electronics technology with applications in the design and development of both civilian space systems and ballistic missile systems.

Analyst Comment: The increased reporting during FY12 concerning South and Central Asia interest in space systems technologies with applications to missile design coincides with a general increase in industry reporting showing foreign interest in missile technologies, as described in the special focus area of this publication. Ballistic missile systems are dependent on complex systems that require years of R&D to perfect. South and Central Asia entities probably view the acquisition of U.S. information and technology as necessary to more quickly address gaps in indigenous technology capability. (Confidence Level: Moderate)

OUTLOOK

South and Central Asia regimes' intentions to increase their military capabilities combined with poor economic conditions and squeezed defense budgets very likely mean a continuation of the perceived need to acquire foreign, in particular U.S., technology. Strained relationships with neighbors, internal counterinsurgency operations, and uncertainty over the future situation in Afghanistan all contribute to South and Central Asia motivations to field modern and well-equipped militaries. Especially given evolving South and Central Asia relationships with the United States, DSS assesses that regional entities will very likely continue their attempted collection against and acquisition of U.S. information and technology. (Confidence Level: Moderate)

South and Central Asia governments continue to place significant emphasis on the development of a domestic industrial capability to meet military needs for technology and system development. To support this goal, and until they achieve it, South and Central Asia entities will almost certainly continue their efforts to gain access to sensitive or classified U.S. information and technology. (Confidence Level: Moderate)

In the near term, South and Central Asia entities will almost certainly view reverse-engineering of U.S.-manufactured components or subsystems as an attractive option to make up for indigenous technology shortfalls. Those regimes concerned about their relations with the United States will be unlikely to engage in diplomatically provocative illicit technology acquisition efforts, but some of their government entities will probably discreetly attempt to reverse-engineer any technology and products they legitimately acquire. (Confidence Level: Moderate)

Conversely, any regimes less concerned about the state of their relations with the United States will likely become more transparent about the closeness of their military and economic support relationships with other powers. South and Central Asia regimes that are in partnerships with foreign powers are likely to aggressively exploit any remaining interactions with U.S. cleared contractors and share any U.S. information and technology to which either partner gains access. (Confidence Level: Moderate)

Notwithstanding these issues, the commercial drive for profit remains a significant factor. Some U.S. companies, including cleared contractors, are increasing their engagement with South and Central Asia business entities, and the relationship between cleared industry and some South and Central Asia defense customers continues to mature. DSS anticipates that a change in perception will likely occur: that cleared contractor personnel, especially those in business and sales development, will increasingly view South and Central Asia-originating inquiries received through these established avenues as legitimate—and that such contacts will increasingly go unreported to DSS. DSS assesses that these contacts will present a risk that illicit entities, including those working on behalf of countries in other regions, will successfully target cleared contractors involved in the South and Central Asia defense market. (Confidence Level: Moderate)

As some cleared contractors become more involved in South and Central Asia defense markets, interaction increases between them and South and Central Asia entities. This will almost certainly present more opportunities for targeted use of the exploitation of relationships MO in attempts to acquire sensitive or classified information and technology. South and Central Asia customers are likely to increasingly attempt to leverage business relationships, exploiting them so as to maximize their access to U.S. technology. The technology acquisition threat from South and Central Asia collection efforts will probably begin to mirror those posed by collectors linked to other countries with which the United States regularly shares military technology. (Confidence Level: Moderate)

Cleared contractors associated with academic institutions are a unique subset of cleared industry that will almost certainly continue to receive substantial numbers of academic solicitations from South and Central Asia entities. This is likely to keep the government-affiliated category at or near the top in frequency of reporting from cleared industry related to South and Central Asia. (Confidence Level: High)

DSS assesses it as very likely that South and Central Asia technology collection efforts will also continue to heavily employ commercial entities as procurement agents attempting to acquire U.S. information and technology. These entities will almost certainly continue to prefer the AAT and RFI MOs. Outside of suspicious network activity, these are generally the lowest-risk, highest-gain methods for attempting to collect U.S. information and technology. While most such contacts will very likely come from legitimate entities and reflect official government requirements, some will probably derive from disreputable actors attempting to obscure the ultimate end user. (Confidence Level: High)

The U.S. technologies that South and Central Asia entities seek to acquire for numerous ongoing and unrelated defense projects span the technological spectrum. DSS expects this breadth of interest will probably continue throughout FY13. (Confidence Level: Moderate)

DSS assesses that South and Central Asia collection efforts are likely to continue against U.S. electronics and LO&S systems to support force-wide modernization requirements and upgrades. A number of the FY12 South and Central Asia-connected reports attributed to these categories concerned requests for enabling technologies, which would very likely have been employed in air defense systems. We judge that this pattern will probably continue as South and Central Asia regimes seek to protect their countries' air defense networks from the likelihood of air strikes by hostile neighbors. (Confidence Level: Moderate)

As South and Central Asia regimes begin the process of shifting to space-based systems, DSS anticipates that South and Central Asia-connected targeting of cleared industry will increasingly focus on technologies with space security and counter-space applications. (Confidence Level: Moderate)

CASE STUDY

During FY12, an individual representing a commercial South and Central Asia procurement company emailed a cleared contractor on behalf of an unidentified end user in the region with a request for a quote for an electronics system with signals intelligence applications.

The cleared contractor informed the procurement company that the requested product fell under the provisions of the International Traffic in Arms Regulations (i.e., ITAR) and the prospective purchasers would need to provide additional end-user and end-use information before sales discussions could proceed.

The procurement agent informed the cleared contractor that the end use was for an airborne platform, but did not identify the end user. The procurement agent then requested that the cleared contractor identify specific programs that used the electronics system in question.

The cleared contractor declined to provide this information to the procurement agent, and again requested end-user information. In response, the agent requested information on whether or not a specific third-country program used the electronics system, stating that such information was necessary to secure an order from the end user, which he claimed had to take place before he could secure an export license, whereupon he would identify the end user of the requested system.

There are multiple possible explanations for why the procurement agent was reluctant to identify the end user, not all of which involve illicit intent. It should be noted that a procurement agent's role as middleman between supplier and end user could be jeopardized should a supplier learn the identity of—and decide to deal directly with—an end user.

In response to the procurement agent's question, the cleared contractor stated that the specific third-country program the procurement agent had identified did not use the company's electronics system.

Analyst Comment: This case study illustrates the combined use of the AAT and RFI MOs in attempting to collect information concerning a cleared contractor's involvement with specific programs. DSS was unable to determine whether an actual end user for the requested electronics system existed, or if instead the incident simply represented an attempt to collect information on specific platforms that use the electronics system in question. Regardless of the legitimacy of the initial contact in this case, we assess it is likely that with increasing frequency South and Central Asia entities will use the allure of a potential sale to attempt to acquire sensitive or classified information and technology resident in the U.S. cleared industrial base. (Confidence Level: Moderate)

EUROPE AND EURASIA

9%
of FY12 reporting originated in **EUROPE & EURASIA**

That is a **13%** increase in the number of cases

Commercial entities were responsible for **43%** of the FY12 targeting attributed to **EUROPE & EURASIA**

150% increase in the reported use of **foreign visit**

70% increase in the reported use of **solicitation or marketing services**

16% decrease in reported cases targeting aeronautics

OVERVIEW

In fiscal year 2012 (FY12), entities from Europe and Eurasia made it the fourth most active region in reported collection attempts against U.S. information and technology, compared to third in FY11. Despite this slip in position, industry-reported collection attempts attributed to Europe and Eurasia increased by 13 percent over the preceding year, and the region contains some of the most skillful—and worrisome—collectors targeting U.S. information and technology.

Relatively speaking, Europe and Eurasia is one of the more stable regions of the globe. Nonetheless, it has its share of rivalries, frictions, and geostrategic concerns. It is also home to some of the most successful and aggressive economic competitors to U.S. cleared contractors. Finally, Europe and Eurasia's very stability serves to sharpen efforts to illicitly acquire access to sensitive or classified information and technology resident in the U.S. cleared industrial base, in that European Union and national economic woes combine with popular pressures to constantly exert pressure to maintain adequate defense establishments on leaner and leaner budgets.

Europe and Eurasia is home to almost all the countries once considered "Great Powers," and a number of them still aspire to maintain spheres of influence within the region and elsewhere around the globe. Both active regional alliances and traditional international responsibilities continue to make demands on Europe and Eurasia forces. Such roles require military forces that are relatively large, modern, well-equipped, and flexible, with a quick deployment capability.

Consequently, in FY12 collectors linked to Europe and Eurasia targeted technologies from almost every sector of the Militarily Critical Technologies List. While the most targeted, according to industry reporting, were lasers, optics , and sensors (LO&S); aeronautics; and information systems (IS) technologies, accounting for 15, 13, and 13 percent apiece, the next 43 percent was spread fairly evenly across ten additional technologies.

Commercial entities were the primary reported Europe and Eurasia collectors in FY12, at 43 percent accounting for more than twice the portion of the next most active at 19 percent, which was government-affiliated entities, often research institutions. Individual collectors accounted for 15 percent. The final two affiliations, unknown and government, each accounted for 11 percent of the total. FY12 percentages for all of these affiliations were akin to their FY11 counterparts, as their numbers of reported cases rose in parallel with the overall increase.

Based on industry reporting, the preferred methods of operation (MOs) for Europe and Eurasia-connected collectors were attempted acquisition of technology (AAT) and request for information (RFI), which at 32 and 28 percent respectively combined to account for 60 percent of all reported suspicious contacts ascribed to this region. Solicitation or marketing, foreign visit, academic solicitation, and exploitation of relationships accounted for an additional 11 to five percent of industry reporting apiece.

At the other end of the spectrum, Europe and Eurasia also contains countries whose export-control regimes are less robust. These countries sometimes find themselves used as pass-through sites for illicit technology-collection attempts, whether wittingly or unwittingly.

COLLECTOR AFFILIATIONS

Within Europe and Eurasia, those attempting to collect against sensitive or classified information and technology resident in the U.S. cleared industrial base fall onto a spectrum based on their degree of association and cooperation with government intelligence collection efforts. As noted, commercial entities are by far the most common attempted collectors in industry reporting at 43 percent of the total. However, some commercial companies are openly affiliated with government agencies in procurement networks; other commercial entities work cooperatively with government intelligence services to support national defense requirements; sometimes only an occasional intelligence service surrogate is active within a relatively independent commercial sector; and still other commercial entities that actively seek to collect against sensitive or classified U.S. information and technology do so only to achieve a profitable edge in the commercial marketplace.

As a result, some commercial entities seek U.S. information and technology quite openly; others demonstrate a willingness to engage in borderline-underhanded maneuvers to do so; and still others intend and attempt to out-and-out steal sensitive or classified information and technology whenever they consider it necessary. As an example, industry reported one case of a commercial company from Europe and Eurasia emailing a cleared contractor requesting to purchase secure communications equipment. The discussions included a straight-out "what-if" question to the effect that if delivery was not permitted to the company's country, would it be permitted to a named third country instead.

At the end of the spectrum toward commercial cooperation with intelligence collection, some national intelligence services use commercial cover in collection operations. Such activities include buying technologies and attending trade shows displaying targeted technologies.

Analyst Comment: The preeminence of commercial entities involved in reported suspicious activity from Europe and Eurasia is likely due to a combination of: commercial sector entities attempting to acquire technology for resale; commercial companies fronting for third-country end users; and governments using commercial entities and public tenders to cloak their involvement in the targeting of U.S. technologies. (Confidence Level: Moderate)

The Defense Security Service (DSS) assesses that some militaries within Europe and Eurasia almost certainly rely on the commercial sector to meet their technology needs, leading to the prevalence of commercial collectors in industry reporting on the region. Certain civilian economic sectors undergoing modernization processes probably adhere to military requirements in their pursuit of improved technology. (Confidence Level: High)

Figure 15: Collector Affiliations

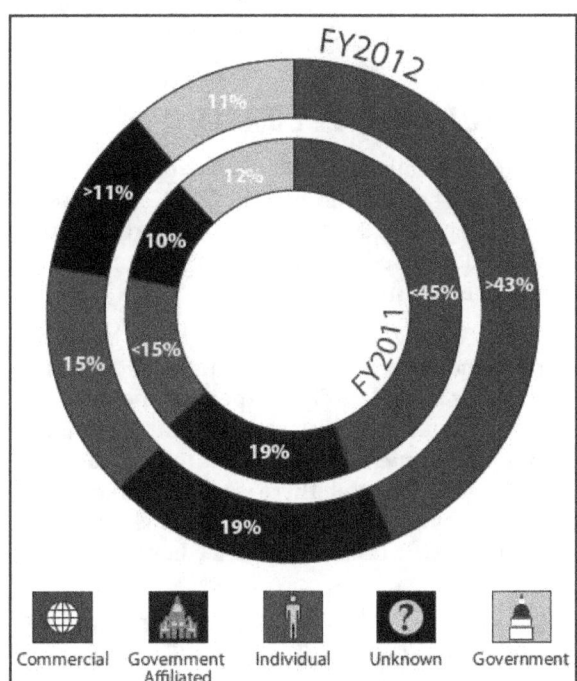

Collection attempts by government-affiliated entities, largely research and educational institutions, remained steady at 19 percent in FY12 industry reporting, representing the second most prominent collector affiliation.

Analyst Comment: Almost all countries in Europe and Eurasia enjoy at least nominally favorable relationships with the United States, including, in some cases, partnering in strategic military agreements. Motivation to maintain favorable military and economic relationships with the United States likely contributes to government-affiliated entities accounting for relatively few reported suspicious contacts. (Confidence Level: Moderate)

Consistent with FY11 data, individual was once again the third most common collector affiliation, at 15 percent.

Analyst Comment: Statistics on government-affiliated and individual collectors often move in opposition to each other: depending on the state of a particular country's relationship with the United States, those acting on behalf of institutional collectors are more or less likely to acknowledge that connection, with a common alternative being to claim independent status. Given the ebbs and flows of relationships between the United States and countries within this varied region, this inverse relationship likely explains the longitudinal stability in these paired statistics, considered together. (Confidence Level: Moderate)

METHODS OF OPERATION

The MO that collectors linked to Europe and Eurasia most commonly used in FY12, according to industry reporting, was AAT. The proportion this category accounted for remained consistent from FY11 at about one-third.

Analyst Comment: It is likely that penny-pinching Europe and Eurasia collecting entities increasingly find the AAT, a relatively inexpensive method, to be useful for initial attempts to gain required technology. If such overt attempts to purchase a product are successful, there is no need to devote the more expensive resources of an intelligence apparatus to the task. (Confidence Level: Moderate)

The second most commonly reported MO in FY12 industry reporting on Europe and Eurasia, at 28 percent, was the RFI. Technology trade shows

are a common venue for RFIs, as collectors target technical specifications of displayed systems or the information in cleared contractor employee presentations. At one 2012 engineering conference, a representative of a Europe and Eurasia research institute approached a cleared contractor employee and asked several questions about a presentation on combustion initiation modeling the employee had given.

Analyst Comment: DSS assesses it as likely that foreign intelligence entities (FIEs) sometimes use such an approach not only to attempt to elicit technology information in the short term but also to build rapport with cleared contractor employees so as to establish a basis for contacting and developing potential recruitment targets in the future. (Confidence Level: Moderate)

DSS assesses that Europe and Eurasia collectors probably consider AAT and RFI to represent low-risk, potentially high-gain methods that are seemingly innocuous and carry low risk of annoying the United States, with which their countries attempt to remain on good terms. (Confidence Level: High)

Figure 16: Methods of Operation

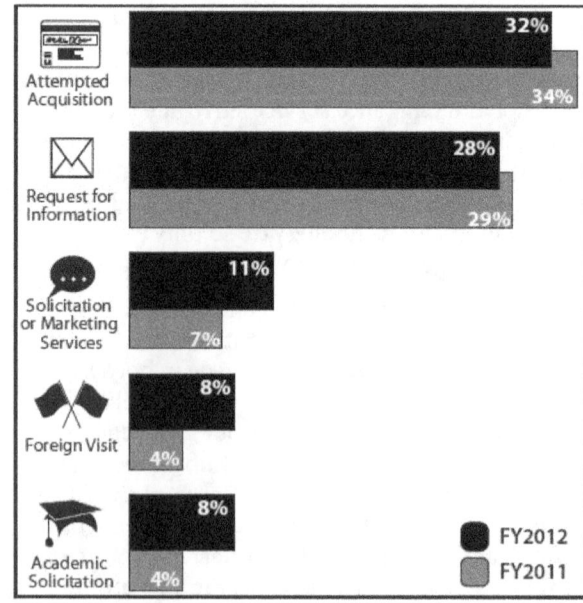

Figure illustrates the top five most reported methods of operation in FY12 compared to the same methods for FY11. None of the other methods totaled more than five percent of reporting.

Solicitation or marketing, foreign visit, and academic solicitation were also commonly used Europe and Eurasia MOs in FY12, accounting for around 10 percent of industry reporting each. The first and third of these represent longer-term methods of arranging relationships with cleared contractors that could become close enough to achieve illicit access to sensitive or classified U.S. information and technology.

The foreign visit collection method holds greater potential both for immediate gains and unpleasant confrontations. But in a situation in which some Europe and Eurasia economies are starting to recover yet budgets are still tight, it is cheaper to allot funding to send abroad a delegation—which could include one or more intelligence officers (IOs)—to visit a cleared contractor in the United States than to develop new technologies independently through long-term research and development (R&D).

Analyst Comment: An IO who is part of a delegation visiting a cleared contractor can use that entrée to identify cleared contractor employees with access to sensitive or classified information and technology for future intelligence service targeting. An IO can also bring concealed eavesdropping devices into contractor offices or perform reconnaissance to determine a facility's vulnerability to technical collection. It is very likely that the presence of an IO in a visiting delegation presents several threats to the security of sensitive or classified information and technology held by a cleared contractor. (Confidence Level: Moderate)

While exploitation of relationships was only the sixth most commonly identified MO in industry reporting related to Europe and Eurasia, accounting for only five percent of the FY12 total, it is an approach conducive to the application of modern social networking tools. Some Europe and Eurasia collectors use social networking sites (SNS) to target cleared contractor employees.

For instance, several cleared contractor employees received suspicious contacts via SNS from Europe and Eurasia entities during FY12. At least three employees of the same cleared contractor received requests to join the social network of a particular individual. The individual may have been a U.S. citizen and claimed to be a program manager for the same cleared contractor. However, the individual had a name that was characteristic of a particular Europe and Eurasia country; public record searches yielded no information verifying the individual in the state in which he claimed to work; and the cleared contractor had no record of an employee with the name the individual claimed.

Analyst Comment: Actual identities of SNS users are difficult to verify. Because of this lack of documentation, it is likely that some SNS accounts represent online personae that IOs use to contact cleared employees. They probably use SNS to contact some U.S. government and cleared contractor personnel just prior to or shortly after visiting Europe and Eurasia. DSS assesses that FIEs from Europe and Eurasia will probably continue using SNS to contact potential intelligence targets because it is a low-risk tactic with the potential to yield significant intelligence gains. (Confidence Level: Moderate)

While suspicious network activity accounted for only three percent of industry-reported FY12 collection attempts from Europe and Eurasia, the region remains a repository for some of the most sophisticated network intrusion capability in the world. Cyber espionage generated in Europe and Eurasia against U.S. cleared contractors represents a current intelligence gap for the U.S. Intelligence Community (IC). However, the IC assesses that the Europe and Eurasia cyber threat to U.S. government networks is wide-ranging, including strong network intrusion skills, advanced cryptographic attack capability, and a software implant capacity that is sometimes deployed through insider access.

Similar to the discussion of collector affiliations, cyber actors operating from Europe and Eurasia also fall onto a spectrum based on their degree of association and cooperation with intelligence services. The covert nature of foreign intelligence agencies' support of cyber criminals complicates IC efforts to attribute hacking activities to governments.

Another complication arises when cyber actors recruit U.S. cleared contractor system administrators for their operations against U.S. cleared contractors.

Analyst Comment: Human-enabled network attacks usually prevent the observation of suspicious indicators normally associated with network attacks. Europe and Eurasia cyber actors very likely possess network intrusion skills that, when combined with compromised system administrators' network access, would enable

the compromise even of computer systems that are hardened against attacks originating over the Internet. (Confidence Level: Moderate)

TARGETED TECHNOLOGIES

In FY12 industry reporting, LO&S, at 15 percent, and aeronautics systems technologies, at 13 percent, were the two most commonly noted, which simply reversed their order from FY11 at comparable percentages. IS technology joined aeronautics in accounting for an additional 13 percent of the total, and the fourth most reported technology was electronics, at nine percent.

These four most common technology categories in the reporting statistics for Europe and Eurasia were the same as for the world overall, albeit in different order. Thus, these represent the most "typical" possible technologies for targeting. In addition, many reported collection attempts came from commercial entities and sought a wide variety of system components with military applications— again providing generalized data that offered few striking statistical and analytical departures, demarcating few Europe and Eurasia developmental initiatives.

Nonetheless, analysis does show some patterns that aid in understanding the reporting data. In industry reporting, Europe and Eurasia collectors demonstrated interest in unmanned aerial vehicles (UAVs); technologies supporting different types of "vision"; and communications technologies. The latter were often components of larger systems contributing to command, control, and communications (C3) capabilities.

Europe and Eurasia provides an example of a military attempting to automate its C3 capabilities by developing multi-level troop-control systems involving these technologies. This particular effort has been ongoing for more than a decade. The developmental program in question has done well at identifying system requirements, but the industrial base is not prepared to build the associated microelectronics in the numbers needed to manufacture and maximize the use of the products. Initial tests of systems under development have yielded little success.

Analyst Comment: Persistent interest in IS technologies likely means that some Europe and Eurasia countries lack effective C3 networks.

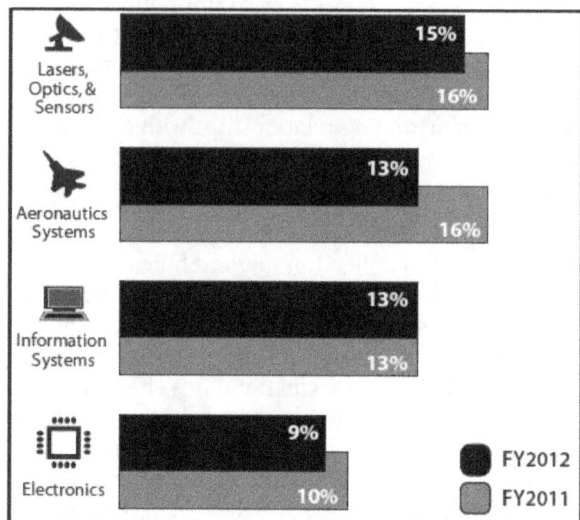

Figure illustrates the top four most reported technology categories in FY12 compared to the same categories for FY11. None of the other categories totaled more than six percent of reporting.

There is an even chance that these unsatisfied requirements signal a deficiency in R&D that is driving collection attempts against IS technologies and, to a lesser extent, electronics. (Confidence Level: Moderate)

Short-wave infrared (SWIR) technology falls within the LO&S technology sector. SWIR technology, sometimes referred to as "reflected infrared" as opposed to "thermal infrared," has a wide range of military and intelligence applications, including ground-, air-, and space-based imaging. For example, a cleared contractor makes SWIR cameras for an enhanced vision system used on board commercial helicopters. A representative of a Europe and Eurasia commercial entity, who claimed to be working with a U.S. business that recommended he contact the contractor, requested SWIR cameras from a cleared contractor.

Even though Europe and Eurasia has its own highly developed defense industries that manufacture world-class military equipment and related products, including aircraft, entities from the region continued to target aeronautics technologies in FY12, often in the form of UAVs and helicopters. During FY12, representatives of a cleared contractor witnessed two Europe and Eurasia nationals photographing and filming aircraft at an air show in Europe. The foreign individuals stated that they wished to purchase a small private aircraft, but observers saw them photographing and filming U.S.-manufactured UAVs and helicopters.

Since Europe and Eurasia produces its own advanced missiles, it is minimally represented in the special focus area of this publication.

OUTLOOK

In general, Europe and Eurasia countries currently seek to build and maintain smaller but well-trained and better-armed militaries. Such evolutions typically involve rearmament, C3 modernization, and the development of joint military operations. To support these purposes, militaries tend to seek more advanced equipment, broadly defined. Europe and Eurasia collection efforts will very likely continue to emphasize the "component technology" categories of LO&S, aeronautics, and IS. (Confidence Level: High)

Much of the transfer of U.S. technology to Europe and Eurasia in support of economic and military modernization will probably continue to occur via legitimate commercial exchanges. However, Europe and Eurasia collectors will probably attempt to draw upon commercial ties to foreign businesses to support additional military technology upgrade requirements via less-aboveboard acquisitions. DSS assesses that commercial will very likely continue as the leading collector affiliation in industry reporting, with government-affiliated remaining a distant second. (Confidence Level: High)

RFIs and AAT have been the top MOs in Europe and Eurasia-connected industry reporting for at least two years. (The trend likely goes back even farther, but prior to FY11 DSS did not differentiate them as separate MOs.) Although not a methodology itself, SNS provides a medium for RFIs, solicitation or marketing, exploitation of relationships, and other MOs. DSS assesses that Europe and Eurasia entities will very likely continue to use RFI and AAT as their preferred MOs. DSS further assesses that some IOs from Europe and Eurasia will probably continue to use SNS to spot and assess cleared contractor employees for potential recruitment, and as a conduit for other MOs. (Confidence Level: Moderate)

The United States and other NATO countries are currently undergoing significant military spending cuts. This will likely give western defense firms, including U.S. cleared contractors, greater incentive to export technology, including to countries in Europe and Eurasia. DSS assesses that collectors from Europe and Eurasia will likely attempt to exploit these conditions to recruit U.S. firms to open branches in the region and thereby strengthen indigenous production capabilities. (Confidence Level: High)

Aggressive collector countries from outside Europe and Eurasia will very likely continue to attempt to use those countries within this region that have less robust technology-protection regimes as pass-through sites via which they can accept delivery of export-controlled technologies. (Confidence Level: High)

CASE STUDY

During FY12, a Europe and Eurasia national approached a cleared contractor's display booth at a cyber-security conference. The individual identified himself as being on an embassy staff in Washington, D.C. He expressed interest in commercial, off-the-shelf products suitable for securing email communications between the embassy and its home country. The cleared contractor employee told the individual that its technology was for U.S. government use only. The cleared contractor employee reported that the individual also attended various sessions on encryption at the symposium.

During the previous year, the same individual had made an earlier, unannounced visit to a separate cleared contractor. He stated that he was checking security vendors in the area to address concerns about hacking of his embassy's network.

Other government agency reporting indicated that the individual was not only an embassy employee, but also an IO.

Analyst Comment: The individual's cover story—that hacking of the embassy network motivated his inquiries—was likely false. Rather, he probably attempted to acquire the technology in question to enable Europe and Eurasia cyber personnel to devise countermeasures to it and to increase the likelihood of gaining access to U.S. networks. (Confidence Level: Moderate)

OTHER REGIONS

Fiscal year 2012 (FY12) was the third consecutive year in which entities from the Western Hemisphere and Africa collectively accounted for a smaller portion of all collection attempts targeting U.S. information and technology reported by cleared industry than in the previous fiscal year. In FY09, entities from these two regions combined accounted for ten percent of collection attempts; in FY12, entities from these regions accounted for just five percent of the reported collection attempts, down further from seven percent in FY11.

In previous years, entities linked to each of these regions increased in number of reported attempts, but at a slower pace than those for the other four regions. In FY12, however, the Defense Security Service (DSS) attributed fewer actual reported collection attempts to entities from the Western Hemisphere (a decrease that was somewhat countered by an increase in attempts attributed to entities from Africa). DSS recorded over a two percent decrease in the number of reported collection attempts originating from the Western Hemisphere in FY12 compared to FY11, in the context of a 60 percent increase world-wide in FY12. Only the Western Hemisphere experienced a reduction in the number of reported collection attempts attributed to regional entities.

Analyst Comment: Collectors from other regions often use front companies or brokers operating in the Western Hemisphere to target U.S. technologies. It is likely that improved attribution of these attempts to the actual region of origin partly accounted for the reduction in the number of reported attempts assigned to the Western Hemisphere. (Confidence Level: High)

The number of collection attempts originating in Africa increased by 94 percent in FY12 compared to FY11. However, in relation to the other regions, its actual increase in number of reported collection attempts was less impressive: it equaled less than two percent of the increase in attempts attributed to entities from East Asia and the Pacific.

In both regions, commercial was the most common collector affiliation. Commercial entities accounted for 46 percent of the reported suspicious contacts from the Western Hemisphere and nearly two-thirds of those from Africa. Based on industry reporting, entities from the Western Hemisphere relied heavily on requests for information (RFI) and attempted acquisition of technology (AAT) when targeting technologies; collectively, these methods accounted for 66 percent of the reported contacts. Similarly, entities from Africa relied on AAT and RFI in 80 percent of their reported contacts.

In FY12 industry reporting, entities from Africa most commonly targeted lasers, optics, and sensors (LO&S). Entities from the Western Hemisphere preferred to target electronics, information systems technology (IS), and LO&S, based on industry reporting. Collection efforts originating in the Western Hemisphere mirrored the reported targeting of entities from East Asia and the Pacific and the Near East in their preference for the same three technologies. This parallelism extended to changes from FY11. In FY11, for contacts originating from the Western Hemisphere, East Asia and the Pacific, and the Near East, electronics was the third, third, and sixth most targeted technology sector, respectively, and IS was the most targeted technology. In FY12 industry reporting for all three regions, electronics became the most targeted technology, IS dropped to the second most common, and LO&S rounded out the top three.

Analyst Comment: Collectors from the Western Hemisphere targeting the same technologies as collectors from the most active collecting regions likely signifies that these entities were acting as the front end of a collection effort originating in another region. These front-end entities likely are either front companies that foreign collectors established specifically to cloak their own identities and those of the end users, or brokers working wittingly or unwittingly with the foreign collectors. (Confidence Level: Moderate)

The number of targeting attempts against U.S. technologies originating from or directed through these two regions will probably continue to increase. However, better attribution of the actual origin of collection attempts would likely cause the number of reported attempts ascribed to Western Hemisphere collectors to increase at a slower pace or decrease slightly, as in FY12. (Confidence Level: Moderate)

CONCLUSION

Technologies resident in U.S. cleared industry remain the target of a wide variety of organized collection efforts originating from a myriad of entities worldwide. This activity may appear to constitute an ill-defined threat. However, over the past few years the Defense Security Service (DSS) and the Intelligence Community (IC)—bolstered by reporting from cleared industry that has improved in both quantity and quality—have been able to gain better fidelity in identifying and quantifying the threat. DSS' analysis of reporting from cleared industry and the IC has better defined the threat by improved identification of the collectors and their origins, methodologies, and targeted technologies.

Reporting from cleared industry in fiscal year 2012 (FY12) largely adhered to the trends of the previous five years. The top two most active collector regions, East Asia and the Pacific and the Near East, remained the same as in FY11. According to reporting from cleared industry, these same two regions have been the most active regions since at least 2007. Similarly, commercial remained the most common affiliation in FY12 and has been since 2007. In the most noticeable change regarding affiliation, government entities became the second most common affiliation. The volume of reported suspicious network activity (SNA) continued to increase, becoming the most common method of operation (MO) used to target U.S. technologies in FY12 industry reporting. The top four most targeted technologies remained the same; however, electronics technology went from being the fourth most targeted technology in FY11 to the second most targeted in FY12.

As mentioned above, East Asia and the Pacific and the Near East have been consistently the most active and second most active regions, respectively, through the last six fiscal years 2007-2012. In FY12, collectively these two regions accounted for 66 percent of reported suspicious contacts. Equally consistent, the Western Hemisphere and Africa have been the fifth and sixth most active regions throughout the same period. During this period, the only variation in the relative level of suspicious

activity from regions has been South and Central Asia and Europe and Eurasia alternating between the third and fourth most active collectors.

During the previous five years, East Asia and the Pacific accounted for between 36 percent and 44 percent of all suspicious contacts during each of the years. In FY12, DSS attributed over half of all suspicious contacts to entities originating in East Asia and the Pacific. Cumulatively over the past six years, East Asia and the Pacific-connected collectors have accounted for 46 percent of all suspicious contacts cleared industry reported.

The East Asia and the Pacific region's share of total reporting represents a larger slice of a larger pie. The number of all SCRs reported to DSS in FY12 represented a 419 percent increase over FY07. However, the number of SCRs attributed to East Asia and the Pacific increased by over 500 percent in FY12 when compared to FY07.

Analyst Comment: East Asia and the Pacific features many areas with a permissive environment in which collectors can operate. Lax export controls and in other cases government tolerance or sanction likely make the region favorable to commercial and individual entities targeting U.S. technologies. Although most technology collected by entities from this region is for end users in this region, it is very likely that a portion of the illicitly acquired technology transits to end users in other regions. (Confidence Level: High)

The Near East remained the second most active collector region according to reporting from cleared industry. In FY12, entities from the Near East conducted 16 percent of cleared industry-reported suspicious contacts. Since FY07, when the Near East accounted for 20 percent of suspicious contacts, cleared industry reporting of suspicious contacts from entities in the Near East has increased by 317 percent.

In FY12, foreign entities identified as commercial made up the most common affiliation in industry reporting on the targeting of sensitive or classified U.S. information and technology. Commercial has been the most common reported affiliation overall since FY06. However, DSS attributed only 29 percent of suspicious activity to commercial entities in FY12, down from a high of 49 percent in FY09. It remained the most common affiliation for entities from Europe and Eurasia, the Western Hemisphere, and Africa. Government-affiliated entities were the most common collectors from the Near East and South and Central Asia.

Government was the second most common affiliation of entities targeting U.S. technologies. In FY11, it was the fourth most common collector affiliation, conducting 17 percent of all suspicious activity. In FY12 government entities accounted for 25 percent of the reported suspicious activity, the second consecutive year in which government entities increased their share of reported suspicious activity. The number of such reports increased by 243 percent from FY10 to FY11, and another 140 percent during FY12. It was the most common affiliation for entities from East Asia and the Pacific. Government entities conducted 41 percent of the suspicious activity attributed to East Asia and the Pacific. These government entities relied heavily on SNA. Government entities from this region accounted for 62 percent of all reported SNA during FY12.

Analyst Comment: It is likely that continued diligence and improved cyber security awareness in cleared industry has contributed to the increase in reporting of SNA. Along with improved recognition of SNA by industry, improved cyber forensics has enhanced DSS, law enforcement, and the IC's ability to attribute SNA incidents to specific entities. This increased reporting and improved attribution has a direct correlation to the increasing number of suspicious contacts attributed to government entities. (Confidence Level: High)

In FY12, for the first time, SNA was the most commonly reported MO overall, at 29 percent of all incidents. The number of such incidents increased by 183 percent in FY12 over FY11, elevating SNA from being the third most frequently reported. As previously stated, entities from East Asia and the Pacific conducted a vast majority of the SNA reported by cleared industry. East Asia and the Pacific-based entities conducted 72 percent of all SNA, with government entities being the most active.

Attempted acquisition of technology (AAT) was the second most common MO in FY12. Although no longer the most common, as it was in FY11,

the actual number of reported incidents of AAT increased by nearly 42 percent in FY12. This was the most prevalent MO among collectors from Europe and Eurasia, South and Central Asia, and Africa. Academic solicitation was the third most reported MO. The number of suspicious contacts using academic solicitation continued to rise in FY12 industry reporting, increasing in number of incidents by 82 percent in FY12 over FY11. Entities from the Near East preferred academic solicitation, using it in 38 percent of the incidents attributed to this region.

The technologies targeted most, based on industry reporting, underwent some minor changes from FY11 to FY12. Information systems technology (IS) remained the most targeted, with electronics technology as the second most targeted, but the difference between the percentages of incidents targeting IS and electronics technology was very small. In contrast, in FY11 electronics technology was the fourth most targeted technology. In FY12, lasers, optics, and sensors (LO&S); aeronautics systems; and materials and processing technologies completed the top five most targeted technologies.

Entities from East Asia and the Pacific, the Near East, and the Western Hemisphere all targeted electronics, IS, and LO&S technology as their top three most commonly targeted technologies. In addition, in FY11 all three most often targeted IS, according to cleared industry reporting.

Analyst Comment: It is likely that the similarity of collection pattern of the Western Hemisphere with the two most prolific collector regions is in part due to proxies and front companies working in the Western Hemisphere providing information and technology to end users in East Asia and the Pacific and the Near East. (Confidence Level: Moderate)

DSS attributed 41 percent of all the reported collection efforts targeting IS to entities from East Asia and the Pacific, and another 20 percent to entities from the Near East. The majority of IS-related attempts sought command, control, communications, computers, and intelligence (i.e., C4I) systems or, most commonly, modeling and simulation (M&S) systems. M&S and analysis software is often used in space systems, but East Asia and the Pacific collectors also sought these technologies for simulation centers and future systems research and development.

Entities from East Asia and the Pacific conducted nearly 44 percent of reported targeting of electronics. A substantial number of East Asia and the Pacific entities' requests for electronics technology targeted rad-hard integrated circuits. These circuits have applications in nuclear weapons, aerospace vehicles, ballistic missiles, and other electronics used in environments subject to radiation. A number of East Asia and the Pacific countries have or intend to have space programs and therefore have a perceived need for rad-hard, space-qualified circuitry. Based on industry reporting, Near East interest in electronics included U.S. communications, electronic warfare, and signals intelligence equipment, including cognitive radios, digital receivers and decoders, demodulators, signal processing components, and direction-finding antennas.

Entities from Europe and Eurasia along with entities from Africa most commonly targeted LO&S technology followed by aeronautic systems technology.

In industry reporting, Europe and Eurasia collectors demonstrated interest in unmanned aerial vehicles (UAVs); technologies supporting different types of "vision"; and communications technologies.

Entities from South and Central Asia targeted a wide range of U.S. information and technology; however, they most frequently targeted electronics followed by LO&S. According to open sources, multiple South and Central Asia governments have launched long-term efforts to develop competitive domestic defense industries capable of meeting the breadth of their countries' military equipment and technology needs. In addition to these broad efforts, multiple South and Central Asia governments are currently undertaking a variety of high-profile, high-tech military research and development programs.

U.S. technologies transferred to entities in other regions, whether via legitimate interchange or illicit collection activities, are subject to subsequent transfer to other end users. Technologies lost to a foreign entity, no matter the circumstances, are likely to show up in the designs and capabilities of other countries. From there they can degrade the capabilities and effectiveness of U.S. military equipment as well as the security and effectiveness of our warfighters.

OUTLOOK

Targeting of information and technology resident in U.S. cleared industry originates from over 100 countries world-wide, from a diverse array of entities, and occurs in many forms. The United States remains the leader in research and development (R&D) of new technology in defense sectors and beyond. This diverse technological leadership places a bull's-eye on U.S. cleared industry. Strategic and economic competitors target U.S. cleared industry in order to reduce time and expense in their R&D of cutting-edge technology. No end to this rampant targeting of U.S. cleared industry is in sight.

Those who attempt to collect U.S. technologies will almost certainly continue to target a wide variety of them, spanning the entire spectrum of the Militarily Critical Technologies List (MCTL). Collectors will very likely target, to some extent, technologies in all 20 MCTL sections, as well as related sensitive and classified information held in cleared industry. Collectors will likely continue to focus greater attention on particular technology sections of the MCTL. Overall, information systems (IS); electronics technology; lasers, optics, and sensors (LO&S); and aeronautics systems will very likely experience the most reported targeting from foreign entities. (Confidence Level: High)

IS technology will very likely remain the most sought after category of technology in industry reporting. The IS category encompasses a wide range of enabling technologies that can provide military and commercial advantage. Collectors will likely continue to target command, control, communications, computers, intelligence, surveillance, and reconnaissance (C4ISR); modeling and simulation software; and advanced radio technologies. Countries with developing space programs will likely target IS and other technologies related to space-based reconnaissance and command and control. (Confidence Level: High)

Microelectronics, specifically radiation-hardened (rad-hard) microelectronics, will very likely remain highly sought after technology, as described in the special focus area of last year's version of this publication, the 2012 Targeting U.S. Technologies: A Trend Analysis of Cleared Industry Reporting. (Confidence Level: High)

Technologies relating to LO&S will almost certainly remain a high priority for collectors targeting cleared industry. This especially applies to collectors linked with countries that are modernizing their early warning and ISR capabilities. (Confidence Level: High)

Countries with established missile infrastructures are well situated to exploit U.S. missile system technology, and entities linked to them will very likely continue to target it. Their targeting will likely span all six major missile subsystems. (Confidence Level: High)

The depletion of natural resources is forcing industrialized nations to seek alternatives to non-replenishable fuels. Countries that depend on fossil fuels for industrial energy production will likely seek alternatives to drilling, transporting, and consuming non-renewable resources that are cost-effective and environmentally friendly. Energy technologies developed by U.S. industry that capture natural energy and convert it to a usable form, and thereby decrease national dependency on fossil fuels, are likely to be increasingly sought after. Such technologies of interest will probably include more efficient batteries, high-temperature superconductors, photovoltaic technology, and equipment for biosynthetic processes. (Confidence Level: Moderate)

The increasing interconnectedness of the world will also very likely drive countries to seek state-of-the-art telecommunications technology, in which the U.S. telecommunications industry is a world leader. Foreign entities attempting to target the U.S. telecommunications industry will probably have two goals. First, they will likely seek entry into a U.S. telecommunications market having a potential 235 million subscribers. Second, they will probably seek out those technologies that are likely to contribute to the long-term evolution of the telecommunications market. Some telecommunications technologies are limited to terrestrial usage, so countries that desire to rid themselves of terrestrial communications infrastructures will probably target space-based telecommunications technologies. (Confidence Level: Moderate)

Entities from East Asia and the Pacific will almost certainly remain the most commonly identified collectors targeting U.S. technologies. These entities will very likely continue to target electronics, IS, LO&S, and aeronautics. Entities from this region will also likely target space system technology as countries and private companies from this region work to further exploit space. Government entities conducting suspicious network activity (SNA) will very likely remain the primary collector affiliation and MO targeting U.S. technologies for entities from this region. (Confidence Level: High)

Collectors originating in the Near East will very likely remain the second most active at targeting U.S. technologies. These collectors will continue to seek technologies and information relating to electronics, IS, and LO&S. In addition, they will likely target technologies related to missile systems and missile defense. Government-affiliated entities will very likely remain the most common affiliation for collectors from the Near East, and academic solicitation will likely remain their primary MO. (Confidence Level: Moderate)

The number of incidents of SNA will almost certainly continue to increase. It will very likely remain the most common method of operation (MO) for entities targeting technology resident in cleared industry. Entities, primarily government entities, from East Asia and the Pacific will conduct most of the SNA targeting cleared industry. (Confidence Level: High)

The direct approaches, such as attempted acquisition of technology (AAT) and request for information (RFI), will likely continue to be common MOs. Collectors from Europe and Eurasia will rely heavily on these approaches. Prior to FY12, collectively these approaches have accounted for over 40 percent of all suspicious contacts. However, this portion of the overall number of suspicious contacts will probably continue to decline as SNA becomes more prevalent and entities continue to exploit academic solicitation. (Confidence Level: Moderate)

The incidents of academic solicitation will almost certainly continue to increase, especially from entities originating in the Near East, East Asia and the Pacific, and South and Central Asia. Academic solicitation will likely continue to be the most common MO of government-affiliated entities from the Near East and South and Central Asia. (Confidence Level: Moderate)

The most common affiliation of collectors has been commercial, including in FY12. It is likely that commercial will remain the most common affiliation; however, the number of incidents conducted by government entities rose sharply in FY12 when compared to FY11. Government could become the most common affiliation of collectors of U.S. technologies in FY13 or FY14. The growth in the number of incidents attributed to government entities has gone hand-in-hand with the rise in the number of incidents of SNA. Government actors originating in East Asia and the Pacific have been and will almost certainly remain prolific collectors of U.S. technology and account for much of the increase in both SNA and the number of incidents the Defense Security Service attributes to government entities. (Confidence Level: Low)

Government-affiliated entities will very likely remain the most common collector affiliation for collectors from the Near East and South and Central Asia. In the Near East, the government-affiliated entities will very likely continue to be government-sponsored universities and research facilities. Commercial entities collecting on the behest of governments to meet specific government requirements will probably conduct most of the collection attributed to government-affiliated collectors originating in South and Central Asia. (Confidence Level: Moderate)

Persistent and pervasive foreign collection attempts to obtain illegal or unauthorized access to sensitive or classified information and technology resident in the U.S. cleared industrial base will almost certainly continue unabated for the future. Foreign intelligence entities' MOs may evolve and the specific technologies they target may change, but the constancy and aggressiveness of the campaign of attempted collection will almost certainly not subside. (Confidence Level: High)

REGIONAL BREAKDOWN

Africa	East Asia and the Pacific	Europe and Eurasia	Near East	South and Central Asia	Western Hemisphere
Angola	Australia	Albania	Algeria	Afghanistan	Antigua and Barbuda
Benin	Brunei	Andorra	Bahrain	Bangladesh	Argentina
Botswana	Burma	Armenia	Egypt	Bhutan	Aruba
Burkina Faso	Cambodia	Austria	Iran	India	Bahamas, The
Burundi	China	Azerbaijan	Iraq	Kazakhstan	Barbados
Cameroon	Fiji	Belarus	Israel	Kyrgyzstan	Belize
Cape Verde	Indonesia	Belgium	Jordan	Maldives	Bermuda
Central African Republic	Japan	Bosnia and Herzegovina	Kuwait	Nepal	Bolivia
Chad	Kiribati	Bulgaria	Lebanon	Pakistan	Brazil
Comoros	Korea, North	Croatia	Libya	Sri Lanka	Canada
Congo, Democratic Republic of the	Korea, South	Cyprus	Morocco	Tajikistan	Cayman Islands
Congo, Republic of the	Laos	Czech Republic	Oman	Turkmenistan	Chile
Cote d'Ivoire	Malaysia	Denmark	Palestinian Territories	Uzbekistan	Colombia
Djibouti	Marshall Islands	Estonia	Qatar		Costa Rica
Equatorial Guinea	Micronesia, Federated States of	Finland	Saudi Arabia		Cuba
Eritrea	Mongolia	France	Syria		Curacao
Ethiopia	Nauru	Georgia	Tunisia		Dominica
Gabon	New Zealand	Germany	United Arab Emirates		Dominican Republic
Gambia, The	Palau	Greece	Yemen		Ecuador
Ghana	Papua New Guinea	Holy See			El Salvador
Guinea	Philippines	Hungary			Grenada
Guinea-Bissau	Samoa	Iceland			Guatemala
Kenya	Singapore	Ireland			Guyana
Lesotho	Solomon Islands	Italy			Haiti
Liberia	Taiwan	Kosovo			Honduras
Madagascar	Thailand	Latvia			Jamaica
Malawi	Timor-Leste	Liechtenstein			Mexico
Mali	Tonga	Lithuania			Nicaragua
Mauritania	Tuvalu	Luxembourg			Panama
Mauritius	Vanuatu	Macedonia			Paraguay
Mozambique	Vietnam	Malta			Peru
Namibia		Moldova			St. Kitts and Nevis
Niger		Monaco			St. Lucia
Nigeria		Montenegro			St. Maarten
Rwanda		Netherlands			St. Vincent and the Grenadines
Sao Tome and Principe		Norway			Suriname
Senegal		Poland			Trinidad and Tobago
Seychelles		Portugal			United States
Sierra Leone		Romania			Uruguay
Somalia		Russia			Venezuela
South Africa		San Marino			
South Sudan		Serbia			
Sudan		Slovakia			
Swaziland		Slovenia			
Tanzania		Spain			
Togo		Sweden			
Uganda		Switzerland			
Zambia		Turkey			
Zimbabwe		Ukraine			
		United Kingdom